BROKEN WATERS SING

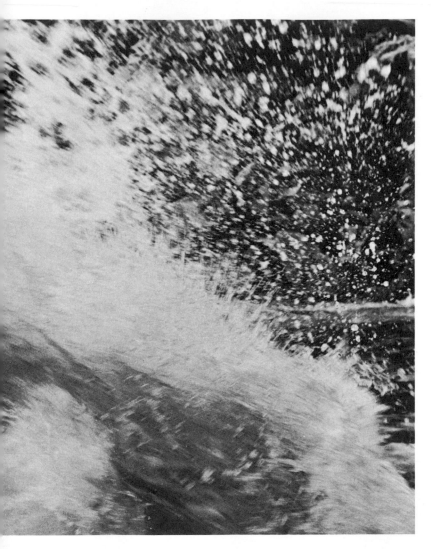

BROKEN WATERS SING

REDISCOVERING TWO GREAT RIVERS OF THE WEST

BY GAYLORD STAVELEY

A SPORTS ILLUSTRATED BOOK

Little, Brown and Company — Boston - Toronto

LIBRARY OF CONGRESS CATALOG CARD NO. 78–149469

FIRST EDITION

т06/71

Sports Illustrated Books
are published by
Little, Brown and Company
in association with
Sports Illustrated Magazine

Published simultaneously in Canada

by Little, Brown & Company (Canada) Limited

PRINTED IN THE UNITED STATES OF AMERICA

*For those who understand
the inseparability of beauty and danger*

I would not choose to have the stream of life
from first to last all calm and rippleless,
For broken waters sing a lovely song
and valor grows on obstacles o'ercome.

—author unknown

CONTENTS

Contents

ILLUSTRATIONS

Illustrations

The Green and Colorado rivers are born in the mountains of Wyoming, Utah, and Colorado. Their headwaters are in the Rockies, the Wind Rivers, the Uintas, and the Wasatches, craggy ranges of 13,000 to 14,000 foot peaks that border their basin. Their canyon-locked course descends 1,400 miles from the mountains to the sea, with only a few points of access.

A hundred years ago virtually nothing was known about the rivers except what men could see at a few places they could reach to cross them, and tales told by Mountain Men who had trapped beaver in some parts of the canyons. Maps of the day bore a large blank area there, and the notation "unexplored." Then came Major Powell.

In 1869 John Wesley Powell was thirty-six, a veteran of the Civil War, and a self-taught scientist who became a professor of geology, and museum curator for the Illinois Natural History Society. In 1868 he had led an exploratory group into the Rocky Mountains west of Denver, and had gone farther west to encounter the chasm of the Green. While there, Major Powell made up his mind that he would explore the great rivers and their canyons.

The 1869 Powell expedition ran more than one thousand miles of rapids-filled river, from Wyoming to Nevada, in small boats. Powell methodically measured and observed the "great unknown" that the rivers and their canyons comprised. With his findings, the map makers were able to erase the word "unexplored" from a last large portion of the American West, and begin showing the course of the great rivers Powell and his men had traveled.

Segments of Powell's route have been run by boat many times in the century since 1869, for varying motives. Commercial running of the Green and Colorado for recreation and sport is only about thirty years old. Historically, like the original voyage of discovery, it was done in wooden rowboats, but the advent of war-surplus pontoons changed the picture by 1950. In 1969 my Nevills Cataract Boats were the only passenger-carrying rowboats still in regular use on Powell's rivers. It seemed appropriate that they should retrace the route he had opened exactly a century before. Ours became a voyage of discovery, too.

Every serious devotee of River has probably read

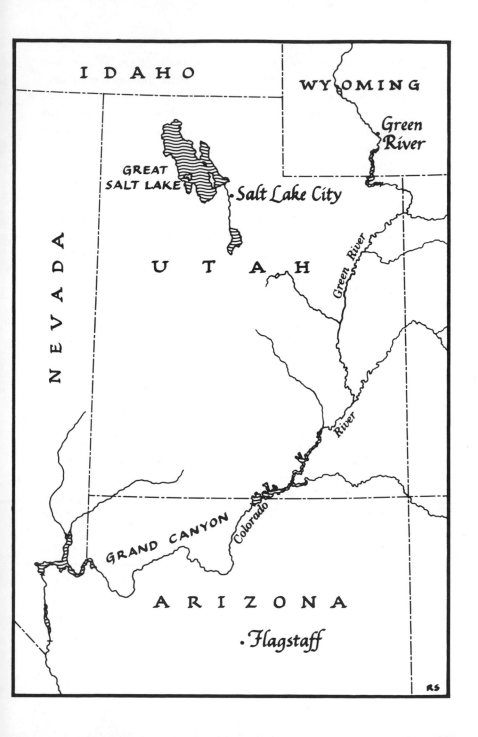

Powell's *Exploration of the Colorado River,* Frederick Dellenbaugh's *A Canyon Voyage,* the account of a second Powell expedition, as well as Julius Stone's *Canyon Country,* an articulate journal of a 1909 expedition, and Robert Brewster Stanton's *Down the Colorado,* about the 1889 railroad survey expedition.

Dellenbaugh's book, to some extent, and the 1914 account of Kolb's expedition *Through the Grand Canyon from Wyoming to Mexico* have been to my knowledge the only books published in which the oarsman himself tells the story of his day-to-day relationship with the Green and Colorado Rivers. A passenger on a river trip need not understand rapids or rowing to be enthralled by the experience, but he sees through different eyes than one charged with the reading of the river, the care of a little boat, and the safety of its human cargo, and he will write of different things than will the oarsman.

It is one of the sorrows of America's environmental awakening that while the two incomparable rivers run through almost continual parkland, they themselves have been unprotected. Within the equivalent of two lifetimes, the Green and Colorado have been discovered, dammed, and dirtied. All of this has happened since Powell's voyage, and much of it since Kolb's. There is still great beauty and high adventure to be found in the canyons. But because there may not always be, the end of the "Powell century" seemed like a good time for another book by a boatman.

The story nearly wrote itself, and if there are errors, I hope they are those of omission, not embellishment. Neither the expedition nor this account of it would have been possible without the help of those who went down the river, passengers and oarsmen alike, or those understanding few who afterward appointed themselves sentinels of my solitude so that I could write. I am especially indebted to Joan Staveley, my wife, Don Staveley, my father, to Verleau and Rose Lee Norton, who put the manuscript into finished form, to Margaret Eiseman and Dr. Michael E. Wiedman, who provided the pictures. And to Peter Wensberg of Boston, who, after I had shown him the canyons of the Colorado, guided me through the canyons of the publishing world.

G.S.

Flagstaff, Arizona

ONE

THE GREEN

DOWN A PAPER RIVER

There were six of us, three in each Cataract Boat, floating down the beautiful canyon of the Green River in northeastern Utah. We had launched the previous day at the toe of Flaming Gorge Dam. Forty-three more days and almost seven hundred miles of river lay ahead; river that in that distance dropped more than four thousand feet, through hundreds of riffles and rapids. Farther downstream we'd add other boats and passengers but now there were just we six who had run parts of the river before and wanted to run all of the yet unimpounded river the old historic way, in rowboats, in the hundredth anniversary year of the Powell expedition.

For a canyoneer to be content with having run part of a runnable river is no more possible than for a mountaineer to settle for part of a climbable peak. During thirteen years as a professional guide on the publicized, romanticized sections of the Colorado, Glen Canyon and Grand Canyon, I'd known that one day I would have to find a way to run the rest of the river, the canyons of the Green and upper Colorado. The centennial year seemed an appropriate time both for following the route of Powell's discovery — what remains of it — and for writing something about the river and canyons as they are now, only one hundred years after becoming mappable territory.

The Eisemans and I had begun talking about the expedition four years before. Fred is a day school science department chairman with tremendous zest for the outdoors and the outdoor sciences, especially geology. His wife Margaret — "Maggie," on the river — is an exceptionally good amateur photographer with a love of, and so a keen eye for, natural beauty. Fred and Maggie had run the San Juan, Glen Canyon, and Grand Canyon with me in Cataract Boats for ten summers. Fred is a fine oarsman and could, I knew, take over to lead the expedition on through if anything happened to me; in addition he had volunteered to manage all the meals so that I would have time for other details. He was rowing the *Camscott*, named for my two sons, who would join us later for their first trip through Grand Canyon.

My boat was the *Norm*, the lead boat for all of the expeditions. It was named for Norman Nevills, who pioneered fast-water trips as a passenger-carrying enterprise on the Colorado River in the 1930's. With me was

Doug Reiner, an aerospace engineer who once captained the Wisconsin rowing team. Doug had come to row a third boat; last-minute cancellations had dictated leaving it behind for the first part of the trip but he'd come anyway because he wanted to run the Green, boat or no boat, and would share my rowing until other boats were added. The fifth and sixth members were Sam Carter and Yoshinao Oka, passengers on this leg of the trip, and seasoned river travelers whose interests coincided with ours.

That morning there were four yellow goslings bobbing on the wavelets of the river just ahead of my boat. The current carried us along together and they wobbled erratically trying to make baby feet pull them downstream faster. A Canada goose, wild and beautiful, was swimming efficiently ahead of them, her distance from us a nervous compromise between concern for her babies and fear for her own safety. Her periodic honks of alarm seemed to reiterate something like, "this way kids — and hurry!"; the goslings paddled incessantly with wild-eyed glances back.

We had first come upon the goose and goslings the previous day just before making camp. They were in the company of a big drake. In the morning we surprised them all again as we rounded the bend below camp where they had stayed for the night when we stopped crowding them. When the drake saw us he quickly decided to go for a walk and leave mother and children to find their own solution. I'm sure he wanted this maneuver to be appraised as a diversionary tactic, but in a military context it would have been called an Immediate Advance to the Rear. Never has a "casual stroll" into the bushes been so urgently taken without breaking over into a dead run! Mother and children slid quickly into the river and headed downstream. For. two or three miles they were just a few strokes ahead of us. Against the emerald water, with the mild morning sun on their backs, they were animated, yolk-colored fuzz, like soft wind-up toys.

But as time went on I became concerned about them; they had been in the icy water for almost an hour. Mature waterfowl, sheathed in their oily feathers, can float almost indefinitely. Goslings, I had heard somewhere, can waterlog and drown because they don't yet have that protection. Certainly they wouldn't let us push them to the point where, still paddling for their lives, they sank beneath the waves? It would be an ugly way to remember the beginning of the trip.

I felt a special affinity for the babies and their mother, not because they were any more lovable or more fragile than other wildlife, but because it seemed appropriate that we were being led down the Green River by wild geese.

Historically appropriate. In the early 1900's a trapper named Nathan Galloway helped design a new kind of rowboat to run the Green and Colorado rivers from Wyoming to Needles, California. Forty years had then elapsed since Major Powell's voyage of discovery down the two rivers, and in those forty years the course had been rerun a number of times. But each subsequent voyager had emulated the major's use of longboats, heavily framed and planked to withstand the blows of rocks, and rowed bow first downstream by oarsmen facing upstream. Each subsequent voyager found the river as hard on boats as had the major; not only were oarsmen unable to see easily where they were going, but their downstream rowing, added to the speed of the current, gave them incredibly fast closure on hazards ahead.

While trapping along the wild rivers of the West in the 1890's, Galloway had often watched geese as they swam; the way they moved, the way they rode choppy water. A goose's body, it seemed to him, was a good hull, stable and maneuverable. The wetted portion is basically a distended triangle, with the wide breast as the base and the sides curved, and flared outward, as they come back to a point at the tail. The underside is "raked," riding deeper in the water midway along than at either tail or breast. The feet are at this deepest and widest point, where they steer the goose easily. Galloway believed a boat of the same configuration would be a good boat, and he tacked several crude ones together in the 1890's and tried them.

Something else made sense to him, too, and he used it: the Canadian-Alaskan concept of "drifting," letting the river current carry the boat along, the oarsman not rowing, but instead keeping the stern end turned downstream. In that position he could look directly toward rocks or rapids he might be approaching, and he had all the strength of his oar pulls to hold against the current, or move across it.

When Julius Stone called Galloway to Ohio to advise on the design of a new boat that they would take down the rivers, the trapper was able to have his concepts translated, shop-built, into Michigan white pine. The new configuration came to be called the Galloway-Stone hull. Its distinguishing features were a much lighter construction, a stern-first orientation, and a raked bottom that greatly increased pivotal ease and speed. It could be easily managed by one oarsman. The 1909 Galloway-Stone boat has been the prototype for all subsequent successful fast-water rowboats.

And so, drifting down the river stern first in a more

7

modern version of the Galloway design, I felt about the goose and goslings somewhat as the old mariners felt about the albatross, and not wanting to harry them unduly, I began using the extreme edge of the fast water wherever the current swerved to one side, hoping they'd take the opposite side, slip into slower water, and be left behind. It was deliberately bad river running, too close to shore and in shallow water, and I hoped they'd take advantage of the escape opportunity soon because I didn't like my forced course. The big Canada, though, was doing a good job of reading the river, picking the fast chutes and leading her babies down them. She had just one plan: to get downstream. I wondered how she could read the water so well and yet not understand how to avoid the two big white somethings full of people that were chasing her.

The roar of the rapid was barely audible, at first, but she must have heard it as soon as I did, perhaps before. The loudness grew as the current carried us toward it and I thought her honks became a little more urgent, more frequent. She began, now, looking all around and back at her babies more often. The irritating single-mindedness seemed to be gone. I thought I could sense her computing the pluses and minuses: ". . . danger behind / danger ahead / babies soaked, exhausted / sheer cliff right / river widening left: weeds / . . . this has to be it — left bank, let's go kids — now!" and with continuing, urgent honks she led them across the widened river just above the rapid and they dragged themselves into the tall grass at the water's edge. We swept by more intent now on the rapid ahead than on tired, soggy geese.

Down to the point where the goose and goslings had left us, the Green seemed to be an imitation of a wild river, narrow and shallow, with little riffles that hardly constituted rapids; they would rate a one or a two on the river runners' one-to-ten scale of difficulty. But our river profile sheets told us that the situation was to change at Red Creek.

A profile sheet shows the river's declivity toward sea level. It looks like a line graph trending downward. But the trend is not the primary concern; it is any change in the trend. A sharp change of river profile sets off an emotional alarm in an oarsman, just as a sharp change on a sales chart does in a sales manager: it means something is happening. Where the profile line drops abruptly across two or three of the parallel lines of elevations it shows the river is dropping through a notable rapid.

Through the years I had become accustomed to running familiar river by the sight and sound of it, and by images of each bend and each rapid that I had stored away. Navigation of new river by maps and profiles was at the same time rather exciting and rather uncomfortable. We had, on our first day on the Green, run fifteen or twenty drops of four or five feet each that had sharp, rather startling profiles on paper but were found to be just quick little pours over side-canyon boulders, or through narrows where the shores squeezed the river. There had been as yet no long, conventional rapids like those in Marble and Grand canyons, with strong, sleek tongues thrusting down into rolling or thrashing tailwaves. But the profile sheets were promising bigger things ahead; in some places the line of declivity plunged downward even faster than it did forward. Those rapid-profiles looked like the profiles of mountainsides. With the responsibility being mine for determining the hazards involved in running each rapid, and then for putting the *Norm* through first as proof-boat — as a warrant of my certitude — I had to reconcile the charts with the actuality as quickly as possible. When I had learned just how bad the profile of a new river's rapids could look on paper without dictating a stop above to reconnoiter, we'd save many unnecessary landings and still minimize the risk of plunging into something catastrophic. It was rather an anxious margin to tread, but that's part of the exhilaration of rowboat canyoneering.

Just the sound of a rapid tells a lot about its severity and structure; deepness of roar and the distance the sound carries are important clues as to magnitude. The rapid just below the goslings didn't sound big, but the Red Creek profile pitched downward at forty-five degrees through three of the five-foot elevation lines. And the map said we were there. There was no breeze blowing the sound of the rapid away from me as we approached, and, in spite of the alarming profile, I was fairly sure that sight would confirm sound and also say go. I decided to trust the sound as I would on familiar river, and wait a few more seconds to float on around the corner and look into the rapid from near the brink.

Later, a mile farther downstream, a deeper and louder roar told me my preoccupation with the goslings had disrupted my navigational "dead reckoning." That hadn't been Red Creek, just two hundred yards of choppy waves pushing down over a bouldered section at the mouth of a small side canyon. But now, from beyond a hard right bend of the river enough like the one above to be mistaken on the map, the real Red Creek Rapid growled at us, and its voice fitted its profile.

The right bank was sheer cliff to the water. We rowed to the left, where we could tie the long bow lines to scrubby trees. Too anxious to wait for the others, Fred, Doug, and I started down the shore to find a boulder

large enough to stand on for an overview. Red Creek
Rapid was by sound the largest rapid yet, and by pro-
file the most challenging one remaining in the Red
Canyon section of the Green. Ashley Falls, historically
the most challenging, had been covered by Flaming
Gorge Reservoir.

The river turned a hard right and just as it changed
direction Red Creek Canyon joined it from the left. Together they made the arms of a Y, joining to form the
stem. Storm water down Red Creek had carried boulders
of all sizes into the left side of the Green and these lay
all across three-fourths of its course, forcing it against
the sheer cliff on its right. Several large blocks of that
cliff had fallen into the right side of the river. These
clogged the top of a channel that was deep, straight, and
otherwise free of rocks. It was the channel I wanted to
use, but there was no way to get a boat into the head of
it. The river fingered down through boulders, leaving
chutes of varied widths, some wider than our boats, some
too narrow to be considered. It began to look as though
we could run the ones we couldn't enter, and could enter
the ones we couldn't run.

It was late morning now and we were about twelve
miles downstream from Flaming Gorge Dam. A wet rim
along shore told us that the water had recently been
lowered. Not by nature; nature doesn't control the
Green any more, nor hardly do the engineers watching
gauges down inside the dam. A computer three hundred
miles away reads hydropower demands and regulates
flow through the dam's generators. Bureau of Reclamation time-flow charts showed us the computer's daily
release patterns and the times of day and night that
high releases and low releases began arriving at certain
downstream points. According to our flow charts, and

if the computer hadn't changed its daily habits, higher water had begun arriving almost imperceptibly at Red Creek about the same time we had. It looked as though it would make this particular rapid easier to run. When we landed I had set a twig upright in the sand at the very edge to determine exactly what the river was doing, rising to cover the stick, or backing away from it. If it was backing away imperceptibly the run would only get more difficult as we stood reconnoitering.

From our vantage point on the massive red boulder, one course, entered in about the middle of the river, showed promise. "See that vee-shaped hole caused by the two submerged rocks?" Fred and Doug sighted out to where I was pointing. "There's a narrow little tongue of smooth water tending to bridge across it right at the downstream point."

"Yes, I see it," answered Fred. "Do you think a little more water will fill that in?"

"I think it might. I've got a twig stuck in to see whether the river's rising or falling."

"The right leg of that vee is a pretty steep hole," Doug observed. "What happens if you miss and go in there?"

"If we get lost out there, we'd better row for the left side of the vee. The water's moving through it much better and if we get dumped at least we'll be washed on through." It wasn't a comfortable thought, for the frigid water from deep behind the dam had numbed our feet the night before in just the time it took to stand in the shallows and toss duffel ashore. "Okay, then," I continued, "suppose we come in at the top between those two barely submerged rocks straight out from us. Move to the right enough after clearing them to miss that hole thirty yards below, and line up to run the little tongue across the apex of that vee-shaped hole. After that, just work gradually to the right to make sure you go down the right side of that island, downstream. The left side is all shoals."

"Passengers?" asked Fred.

"Have to. I don't think they can get down along shore."

We had been on the rock for perhaps a half hour and during that time the tongue seemed to have widened, and spanned the hole slightly more. We walked up nearer the boats to check the marker twig; it now stood in the water.

"Let's give it a little more time to fill in," I said. With the water rising, our course could only improve. We spent an hour or more exploring, and photographing the rapid, then it was eleven-thirty. We went back to our overlook, saw that the tongue had filled out somewhat, and reviewed the course. There was no need to wait any longer.

Doug coiled the bow line of the *Norm* for me while I fussed nervously with life-jacket straps, and checked safety lines attached to the water canteens and bailing bucket, and checked bow and stern hatch locks, and thought of dozens of other things that could be done to kill time. Procrastinating. Nervous about the first big one, a tight channel with tight entry and the boats at about fourteen hundred pounds each with crew and provisions. He shoved us off and piled in, and I pulled the *Norm* into the current to start lining up. Fred followed with the *Camscott* three or four boatlengths behind. I swung the stern downstream and stood at the oars, hoping for a better look. The two big rocks at the head were easily located; we would enter between them, but everything else was still out of sight below the brink of the rapid. The side-canyon boulders had a damming effect and the river was slowed, nearly stopped above. I dipped over the side and wet the handles of the oars for a stickier grip; there were a few more seconds to locate the holes and chutes we'd seen from shore.

Suddenly I could see bottom and realized the river was shallow above the two entry rocks — we hadn't been able to see that from shore. The *Camscott* was close behind so that I couldn't deliberately ground in the shallows for another look, and if I hung up in the entry chute there was nowhere for Fred to go. I decided to ground in the left side of it, pulled a fast right oar to quarter the bow slightly across the current, held, and let the current move me left, half a boatwidth, enough for them to go by, perhaps. But instead of grounding on the bottom, we brushed the left-hand rock lightly, paused, then floated on over and entered. The next hole was in sight now . . . quarter right . . . a few extra pulls to make up . . . abreast of the top hole . . . vee-shaped hole in sight . . . tongue looking good just ahead . . . on it now — roaring hole on each side . . . clear! Then down through waves in the lower part of the chute with great red rocks close by on each side, then we were pulling, quartered right, sliding past the island and down into calmer water. Fred had reacted immediately to my improvised entry, had made a similar adjustment left past the shallows, and a perfect connection with the elusive little tongue. Now we felt as if we had run a real rapid on the Green.

A few hundred yards downstream there appeared to be a picnic table on the right bank. As we floated closer we could read a government sign — one of those with white lettering on stain-finished planking — designating the RED CREEK BOAT CAMP. We landed; it was noon, and we wouldn't be having lunch from a table just every day on seven hundred miles of river. But more often, it later developed, than we had suspected.

The *Camscott* had happened to become the lunch boat. Maggie and Fred dug through the bow hatch and the compartment under the oarsman's seat and got a wheel of longhorn cheese, two loaves of bread, a bag of cookies, and pickles, peanut butter, and jam. Doug dipped a bailing bucket of icy water from the Green, tore the tops off two packets of drink powder, and stirred up the flavor of the day, which was called Goofy Grape. Youngsters might have fun with such flippant flavors, but I wondered whether a river runner couldn't easily develop a superstition about drinking one called Crash Orange.

Everyone grabbed an armload of jars and packages and we hauled them a few steps up the sloping bank to the big heavy-duty table, which was much closer to the river's edge than it would have been in pre-dam days: the spring runoff from the Uintas and the Wind Rivers into the old undammed Green would have swollen the river at least that high. Now most of these seasonal surges were caught by the ninety mile reservoir behind Flaming Gorge Dam and even the highest possible release from the dam was lower than the spring runoff used to be.

The rim of Red Canyon was about eight hundred feet above us, set back somewhat from the river. A rocky-sandy talus slope slanting down from the skyline cliffs ended just behind our table. Many new seedlings were planted randomly over the slope, interspersed among charred skeletons of juniper and pinion trees. Man-made water pits had been dug around their frail stems. A Bureau of Land Management sign at the toe of the slope explained that the seedlings had been set to reestablish slope protection against erosion, after a careless camp-

Powell's 1869 expedition carried no photographic equipment, but a second Powell voyage two years later did. Expedition photographer E. O. Beaman recorded this scene in June 1871.

The same scene, as we found it ninety-eight years later, on May 27, 1969. A small box elder tree now dominates the foreground and, along the right shore, the sandy beach has been washed away and the tall dead pines have fallen.

er's fire had destroyed the natural growth; it asked users of the camp to water the seedlings.

We drifted on, when lunch was finished and provisions restowed, and within about a mile noticed the height of the canyon walls beginning to recede. Soon afterward, coming around a slight left bend of the gorge, we saw we no longer had a high skyline of cliffs ahead, but of ridges, lower, bordering the river, and much more sky. According to the map, we were entering Browns Park.

The park is a long, narrow valley caused by pronounced sinking of a big block of earth's crust long ago in geologic time. The higher terrain from which it subsided still borders it. On the topographic map these higher forested flanks are tinted green and the valley itself, being lower and unforested, appears uncolored. This gives it the appearance of a long, slightly curved finger pointing through the mountains from northwest to southeast. Some of Browns Park is in Utah, but most of it crosses into Colorado. In width it is about four miles and in length about forty. The Green River follows the valley floor for about twenty miles, but meanders, so that it flows about thirty river miles between entry and exit.

Such pockets in the mountains were once called "holes" and from the time French-Canadian trapper Baptiste Brown settled there about 1827, it has been Brown's Hole. Mountain Men sometimes held their freewheeling spring rendezvous there after a winter of trapping beaver in the high country. It was later a winter range for cattle being trailed north to the transcontinental railroad at Green River City, Wyoming, and a good place to rustle a few head to drive up to Montana or Canada. Outlaw gangs like Butch Cassidy's Wild Bunch holed up there when things got too hot for them.

At the Jarvie Ranch, just ahead of us now on the left bank, the original tenant had been robbed and murdered by two trail bums and his body launched down the river in a rowboat, to be found months later and many miles below.

With the passing of that era, someone had gotten the valley named more respectably Browns Park and de-apostrophied for the map makers. It had become quiet ranch country. For us it was to be an interlude between emergence from Red Canyon and penetration of the Canyon of Lodore.

Suddenly, there was something ahead, just below Mile 274 : a bridge — a very low bridge!

Eight or nine square cages made of logs had been set across in the river at intervals and dumped full of coarse rock for ballast. A roadway of heavy planks had been laid across the tops of these piers from one bank of the river to the other. With less than a hundred yards to go and the current swift, a decision was needed — should we land and look or should we try to run through one of the holes? Five and a half feet for the width of the boat plus a couple of feet for oars pulled in as far as possible; maybe a foot of boat above the waterline plus our height above the seats . . .

But feet and inches came later in retrospect; all I could do as we swept down on the bridge was glance back at the *Camscott* for scale and then mentally project its size against the cruel portals framed between pier and pier, planks and river. I felt that we could make it. "It'll fit," I told Doug and Yosh. "Stay down as we go under." Glancing back at Fred, I couldn't tell whether he was convinced or not; he conceals his nervousness well.

There was one opening that looked slightly larger

than the others and I quartered over, exactly above it. All the fine tuning had to be done before entering; once the boat was under the bridge there could be no dipping of oars. At the last second I swung the blades backward so they'd drag rather than gouge if they hit the piers, we ducked low, and the *Norm* slid into the half-light under the bridge. The coarsely threaded point of a long roadway bolt raked by a few inches over our heads, then sunshine washed all over us again, and we sat up and turned to watch the *Camscott* emerge from the same little tunnel. Later we found a notation of the TAYLOR FLAT BRIDGE in some Bureau of Reclamation river data, but nothing about it on any map.

Two miles below the bridge we encountered a diversion dam of logs and boulders nearly crossing the river, and skirting it by taking a small channel that ran against the left bank, neared an old ranch house fronting on the river. The ranch house itself looked abandoned but a long, wide house trailer was parked beside it. Inside, standing carefully back from her screened door, a woman scrutinized us without coming forward to speak or wave, as we drifted away downstream. She was still watching after a half mile. Then the river turned abruptly from the general eastward course it had taken entering Browns Park and began meandering more southerly. It moved well, at a uniform gradient without rapids or riffles, and we spent the midafternoon hours merely holding the boats in the current with the occasional pull of an oar.

A river that loses elevation at the rate of less than four feet per mile is a slow river, moving downstream very poorly. An oarsman likes a spirited river so that he is always approaching something, always keyed up, and his work at the oars is maneuvering rather than the drudgery of rowing downstream to make up for the slowness of the current. And now we were riding a gradient of eight or nine feet per mile and enjoying it supremely.

Waterfowl were always in sight. At nearly every bend several mallards burst from the water in their characteristic vertical takeoff. Diving ducks of several species flapped and pattered on the river getting flying speed, or tipped quickly beneath the surface. Clusters of ducks rose ahead of us, others already spooked were in the air, and still others were circling to land behind us after we'd passed. Opposite a handsomely kept ranch set in a grove of trees, eighteen big Canada geese suddenly got up in a profusion of alarmed honks. I lunged for my camera, but they were only dots on the ground-glass viewer by the time I got it ready.

A wide ridge, the flank of one of the bordering mountains, projected out into the park ahead of us. The river had run south toward it, then had turned and cut a gorge through it from west to east. Major Powell's 1869 expedition had encountered a profusion of swallows within it, and he later named it Swallow Canyon. We could see brick-red quartzite walls increasing in height and the river starting its turn. Just above, along the right bank, a tall pine and some smaller trees were casting long shadows. I hoped we could camp there; we'd made fifteen miles since morning, run a number of small rapids, one big one, and under a bridge and around a dam. Not knowing Swallow Canyon, I thought at that point it would be nice to save it for the next day. We floated on down, hoping such ideal trees would have an ideal beach beneath them, but they didn't. Just a steep mud bank. Disappointed, but expecting to find a substitute Shangri-La in Swallow Canyon, we slid on by.

On making the turn it became almost immediately obvious there were no Shangri-Las in Swallow Canyon. The gorge was narrow and straight, with very little shoreline, walls in many places dropping straight into the water; what was more, the water had almost stopped moving. I rummaged into the profile sheets. The gradient had leveled at the mouth of the canyon to two feet per mile. And it stayed at about that through the rest of Browns Park. We were being slacked along at about one mile an hour, which meant it would take two hours just to float on through Swallow. The afternoon ride on fast water hadn't been free after all; payment had just been deferred. Now we payed: we rowed.

Not far into the canyon we did find the air filled with swallows, swooping and flitting, catching insects. Forty or fifty hive-shaped mud nests were plastered against a sheltered face of the red rock a few feet above the water. It was a busy little airdrome; all flights began and ended there, and appeared to be strictly local. A swallow-size hole had been built into the face of each nest and its tenant would pop from it headfirst and go for an insect. Judging from the sharp flitting and wheeling, the swallow caught them strictly by outflying them. Then it would return to the nest. Perhaps there were babies inside.

We interrupted our rowing for a few minutes and watched the little air show from a few yards away. It was refreshing to be in a canyon filled with birdsong, but it heightened the absolute lifelessness we found a few hundred yards farther in. There were no more birds, not one, nor any moving thing except our boats. Nature seemed in suspended animation; the air was without motion; the blocky red walls were separated by a narrow river so sluggish against them as to be soundless. The lack of all sound was more noticeable than the subtle outdoor cacophony of the day had been. The canyon seemed an empty narrow way between rows of tall, red buildings from which we were being watched. There was the feeling of sneaking so as not to impose.

Forty-five minutes of rowing put us within sight of open country again. Along the left bank a narrow bench of silt now intervened between the river and the low cliff. We landed and tied there for the night, anxious for that last chance for afternoon shade at the base of the cliff; we had not expected such clear, hot days so far north, even in late May. Fred drove the pointed legs of the long stove irons into the dirt to make the platform for griddle and pots and started a driftwood fire under them, Maggie started a tossed salad, and we others unloaded duffel from the compartments of the boats.

The evening menu called for French onion soup, salad, Salisbury steaks, corn, fruit, and tea. As we ate we could look ahead into the farther reaches of Browns Park; we had floated through about one-fourth of it since early afternoon. The map showed us to be within three miles of the Utah-Colorado state line, but even without the map one could see change ahead. The skyline, several miles distant, was the high wooded edge of the O-wi-yu-kuts Plateau. From our perspective it was a high green rampart, an abrupt beginning of new terrain. It looked like Colorado *should* begin there.

Dinner was finished and pots and griddle scoured and draining when we began to really notice the mosquitoes. There had been a few in the air when we first landed, but with twilight they suddenly seemed numerous — far more numerous. Expecting mosquito trouble somewhere along the way, each of us had included a can or bottle of repellent in his gear; now we pawed through quickly

Fred, Yosh, Doug, and Maggie, at our first night's camp, nine miles below Flaming Gorge Dam, with the north flank of the Uinta Mountains on the horizon.

and located them and sprayed ourselves and each other. It was temporarily helpful, holding most of the little pests a few inches away from ears and necks; they hovered there awaiting another chance. After ten minutes the effectiveness was lost and they'd charge back in for more bites. Another *psst!* from the spray and they'd retreat again and wait. Each of us had his own private cloud of mosquitoes, with replacements hovering at large to fill in for those swatted in battle. It seemed to us that the swallows were hunting insects at the wrong end of the canyon. And we didn't know it yet, but that was only the beginning of two hundred miles of mosquitoes.

I talked a quick tape of the day's events into the little battery-powered recorder, and tried to look through maps and charts, but the mosquitoes were relentless, so I gave up and crawled into the sleeping bag. Pulling it up as far as I could, I draped a shirt over my head, then lay reluctant to move at all for fear of disarranging it. Outside the cloth the whines continued unabated.

In the morning I awoke in nearly the same position. The shirt was still there; I pulled it off and sat up. The sun had risen over Colorado to the east, and most of the mosquitoes were gone. But before leaving, every one of them must have "kissed" me good-bye through the shirt at least once: my forehead was stippled with bite bumps. To my fingers it felt like a landscape covered with ant-hills. The others, too, had received their due shares, on foreheads, temples, wrists, whatever had protruded from their sleeping bags at any time through the night.

Breakfast finished and boats reloaded, we floated on in Browns Park. The river moved somewhat better than it had in Swallow Canyon, but the plotting of our progress between points on the map showed we were still averaging only two miles an hour. About ten-thirty that morning we crossed from Utah into Colorado without any sensation of transition, and almost immediately encountered a riverbank sign bounding the Browns Park National Wildlife Refuge. The wide, marshy bars and gradual, grassy shorelines there were protected territory for pronghorn antelope, and for the ducks and geese we'd been seeing.

By two-thirty we'd reached Mile 255, having made nine miles in five hours. "Anyone interested in rowing down to the head of Lodore instead of spending another night in 'mosquito flats'?" asked Fred. It sounded good; twelve miles more before camp and we'd be through Browns. Slow river and low walls make canyoneers restless.

The Green made its last few miles through the park as a series of meandering loops, one into the next. It became wider and shallower, and increasingly silty with distance below the dam. Its banks were punctuated by big, venerable cottonwoods, often in groves, from which white-faced range cattle stood chewing, watching us pass. On the outside of nearly every bend, where the river ran deepest, heavy creosoted pilings had been driven side by side, standing as a protective shield against undercutting. Big electric pumps were set just behind the pilings with capacious, thirsty intake pipes hanging from them into the river. Most of the pumps were running, humming a powerful high-speed monotone, but there was no one attending them. Except at the outside corners of its great bends, the river ponded along lazily, with deep water hard to find and harder to follow.

A lazy river is unable to continue carrying the load of silt it has picked up in its faster sections or received from side-canyon storms, so it starts dropping this burden particle by particle, farther downstream. Aprons of silt begin building on the bottom, each one increasingly thicker toward its downstream end. These aprons may be as small as a tennis court or as large as a football field. In silty water they are hard to see until the boat is almost upon them, for the water that builds them is the same color. When one of these apron-shaped bars starts to invade a zone of the riverbed, a lot of the bottom water is diverted by it and tends to start moving around it some distance back upstream; this is done so gradually one can scarcely see it starting to go. One way to see it — or sense it — is to look "through" the surface in question, by looking at it, but focusing your eyes longer: beyond. A slight tendency of the surface to be moving one way or the other can then sometimes be detected, but only if it still lies ahead; the movement is so subtle that if you're already on it and moving with it, you can't detect it. Once you've grounded there's nothing to do but get everyone out into the ankle-deep water to lighten the boat, then drag it to the edge of the sand apron, the edge of deeper water.

Lower Browns Park had the kind of river I had run in Glen Canyon, on the Colorado, and had finally taught myself to read fairly well at the cost of a lot of dragging and embarrassment. Our old barb was: "If you're going for a walk, why take a boat?" Both the *Norm* and the *Camscott* were heavy with provisions, drawing about ten inches of water, and it had been a few years since Glen Canyon. I grounded us two or three times that afternoon.

Just above Mile 251 deep water ran against the left bank and a new red frame building had been built at the river's edge. As we rested the oars a moment to study it, a station wagon appeared from somewhere beyond, swung close, and headed up a dirt road on the brim of the bank. An attractive woman was driving, dressed as though she were going shopping, or to a club meeting, either of which it was hard to imagine being found within a hundred miles. She tossed us a friendly wave, casually as though Cataract Boats came down the river regularly, and we felt less like intruders than we had when studied warily from behind a screened door. The car continued on up the road, leaving a twist of dust.

In addition to using river plan and profile sheets, we had been tracing our progress through the surrounding countryside on a series of topographic maps published by the United States Geological Survey. These show the nature of the terrain for many miles adjoining the river. Contour lines and colors indicate valley, upland, and mountain, and cultural features such as roads and buildings are included. Our position late that afternoon was on the Lodore School Quadrangle, named for the map's most apparent feature. The school was seen to be a boxy, unpainted wooden building on a low ridge east of the river. Facing the Green, along the length of the near side, was a row of many-paned, wooden-sashed windows and it was easy to imagine that rows of desks ran along just inside the windows and that boys and girls inside often found their gaze wandering out to the river and the big cottonwood trees. The cedar-shingle roof rose perhaps two stories above the ground and had the sharp pitch given to roofs in snow country.

Vermilion Creek runs tributary to the Green near Lodore School, having come half the length of Browns Park from the east to join; together the two drain the great valley. This union of drainages having been made, the Green immediately heads south and, with very little meandering, leaves Browns Park. According to some geologists, Fred says, there is evidence that the Green

*As we floated the meandering river through lower Browns
Park, the Gates of Lodore were nearly always in sight to
remind us that Disaster Falls, Triplet Falls, and Hell's
Half Mile were only a few hours ahead.*

long ago flowed on eastward from this point, in the watercourse through which Vermilion Creek now flows west. Then over a great period of time a little drainage retrospectively named Lodore Creek cut its way up from the south and captured the Green, tapping it off to the south. Eventually the abandoned eastern section reversed its drainage direction and became Vermilion Creek, a tributary of the rerouted river. Looking at the broad valley along which we had come, and seeing how it continued about the same width and depth to the east, it was easy to accept the idea that the same stream did once carve it all.

Just downstream from the mouth of Vermilion, a great green mountain could be seen rising from the edge of the park, its rounded sides thickly wooded. It was pleasantly symmetrical, refreshingly verdant, one of the easternmost of the high Uinta range. But for two thousand feet, from its summit to its foot, there was a narrow, vertical gash of raw Precambrian rock. Into that gash, between those rocky Gates of Lodore, ran our river.

Our only spoken contact with anyone in Browns Park was in the last mile. Two young men were hand-pumping a pontoon at the Lodore Ranger Station. As we floated by we exchanged howdys with them. Then we had gone by and the Gates of Lodore were dead ahead: mammoth green jaws gaped as if displaying their harsh equipment and waiting to use it on us. We decided to float just inside, and camp.

THE CANYON OF LODORE

It's exciting, and usually disturbing, to study maps of a river canyon that you haven't run before. But on a map nothing is real. A region of mountains and canyons, hills and valleys, trees and grass appears on a flat map sheet as a myriad of brown contour lines drawn through places of equal elevation, farther apart where the country is gentle, crowded against each other where they depict abruptness such as the sheer walls of a gorge. A river is mapped as a contorted double line a quarter of an inch wide. One may know that the river will have fast chutes, rocky shoals, thrashing waves, jutting bedrock ledges, and even quiet untracked beaches. But these are not real until one sees them. A map tells everything, and yet nothing.

Imagination usually rushes in to fill the void between map and actuality. Bits of historic accounts of Powell's, Stone's, Kolb's, or Nevills's rowboat battles with this rapid or that rapid seep in to fill the space, if any, left unfilled by imagination. The mind rounds up all the scares and anxieties connected with other times on other rivers and replays them as supposition of what could happen to you on this new piece of river. It tries to construe and solve the problems and hazards in advance; being unable to do either, it becomes even more anxious. The cycle is almost self-perpetuating. Yet it's nothing more than a kind of stage fright and the only remedy for it is to get out onto the river. Once you're at the oars, each of the river's challenges becomes tangible and individual: a rock ahead on the right — go left; an eddy ahead — go around; an overly loud rapid below — land and go look.

I lay in my sleeping bag just inside the Gates of Lodore. Morning had come; Fred and Doug were already up that I could see, and perhaps the others, too. Coffee was brewing. I was pretending to doze, thinking about Lodore, mostly in terms of numbers:

"Drops 260 feet in 18 miles, 220 of them in 10. Overall 14 or 15 feet per mile; 22 per mile for the 10. Drop of 8 a mile's a lot. Fast drop starts Mile 239, ends 229. Disaster Falls starts Mile 237, drops 37 . . . Triplet Falls starts Mile 232½, drops 14 . . . Hell's Half Mile starts Mile 232, drops 30 . . . tonight we should be . . ."

"Look!" said Fred suddenly, facing back upstream.

"What's happening?" reacted Doug.

"Look!" repeated Fred, looking to make sure he had

everyone's attention. "The morn, in russet mantle clad, walks o'er the dew of yon high eastward hill," and then grinned and added, *"Hamlet,* act one, scene one."

With that a better keynote to the morning than my Lodore reverie, I dragged my trousers in with me, wrestled into them, and then fought the balky sleeping bag zipper halfway down. My commotion in getting out disturbed a few mosquitoes, who had apparently spent the night trying to jab through the thickness of the bag, and they followed me down toward the fire. Back up the bank against the limbs of a box elder tree, Maggie's framed mosquito net soon started to jostle around, and before long she had joined us.

Early sunlight was directly against the east-facing cliff overlooking our little sandbar. The glassy surface of the river had no color of its own, but was reflecting the fiery color of the sun-struck red quartzite. The air was tart, but not cold. Spirits were good; soon we'd translate some of the Lodore map into rocks and waves.

By eight-thirty we were loaded and under way, with about four miles of smooth water before the first drop shown on the profile sheet. The canyon was attractive; friendly, with almost vertical cliffs buttressed at the bottom by broken rock that had fallen from them. In a few places one or two blocks the size of our boats had bounced on out into the river, but it ran quietly around them. Box elder and juniper trees two or three times a man's height edged the banks along the water, with single trees here and there on the talus slopes. Birdsongs of many kinds came from the trees and rocks. The bottom of the canyon was still in shade.

We ran Lodore's first rapid, a short, fairly sharp drop with a big rock splitting its tailwaves, by standing at the oars and studying it as we approached. At Mile 241 we ran a longer rapid, a shallow C bending around to the right, and then landed and walked back to explore a log cabin we'd been watching for. The notation on the map said WADE AND CURTIS CABIN. Two early Browns Park settlers had once tried to start a modest guest ranch there in the upper reaches of the Canyon of Lodore. Their cabin was of sawed logs eight or ten inches in diameter, about eighteen feet long on the sides and twelve feet across the ends. Apparently it had since been reconstructed, as it had a poured concrete foundation and smooth planked floor. Modern picnic tables nearby seemed further evidence of its being fixed up as a tourist stop. And there was a profusion of outhouses: three set randomly around the cabin and at least one more in the trees closer to the river. Between the various members of the group, we discovered that none of them was marked specifically for men, or for women. We couldn't decide whether having more than the customary two, one for each gender, had overly complicated the matter, or overly simplified it.

An old waterline of two-inch diameter steel pipe angled down to the back of the cabin from a patch of lush greenery above, at the top of the talus slope. Bypassing the log cabin, the line trailed off into the tangled juniper-tamarisk growth nearby. Sam, tracing it, discovered what had probably been the original Wade and Curtis cabin. The sawed boards that were once walls and roof had been nailed up without benefit of joists or rafters to stiffen and support them, and time had taken them apart. They had collapsed onto the floor and furnishings, and the head and foot of an iron bed were draped with fallen, weathered boards. It took up most of one side of the floor area. Opposite, near where the door had been, the lids and plates and doors of an old Excelsior stove occupied a lot of the remaining space. The cabin had perhaps been ten feet square. It looked as though it might have accommodated either Wade or Curtis, but not the two together.

Precisely at Mile 239, the canyon of the Green makes a pronounced jog to the right. At that point, as if it had been waiting for such a signal to deviate from four rather straight miles, the river starts its fast charge of ten miles downhill at twenty-two feet per mile. Within those ten miles are Lodore's "big three"; Disaster Falls, Triplet Falls, and Hell's Half Mile. Within a mile after leaving the cabin, I could see the left-hand canyon wall curving across in front of us, and knew that was the threshold.

During the fifteen minutes it took to float down, I scanned both sides of the river rather nervously for shore where we could land if the river didn't sound right from above the corner. As we neared each usable little bit of riverbank, I'd hold it in abeyance and look for a substitute farther down, so we wouldn't pass the last possible one without any place to stop if need be. But there seemed to be another one always just ahead on one side or the other, and the river still was not bellowing a deep-throated roar at us that would suggest ragged waves and deep thrashing holes.

At the corner it still sounded good. A gray sign came into view, just upstream from a left-bank side canyon:

TRAILER DRAW ESCAPE ROUTE
DANGEROUS RAPIDS AHEAD

Then we could see the abrupt beginning of the drop ahead. The pitch of the canyon floor steepened noticeably. Everything still said "go."

The river immediately narrowed, slipped over the brink, and accelerated from four miles an hour to nine or ten. Still, there appeared to be plenty of gentle shoreline with eddies of slower water where boats could be eased in and landed without banging along the bank. For two wonderful miles the Green was a continuous sluice with no individual rapids and only a few midstream rocks, all easily detected well in advance. We raced along exhilarated, quartering, pulling, holding, always maneuvering, but never with the sensation of plunging irretrievably into something we couldn't handle.

After several minutes at such velocity I was sure we must be nearing Mile 237: Disaster Falls. A slightly louder roar ahead, overpowering the rattle of water around us, seemed to say this was so, but standing at the oars I could see only a small rapid, all of it visible, with nothing below but a resumption of swift water. It obviously wasn't Disaster Falls, or if part of it, then only an easy first drop. We could easily run that and land below before the river turned and went out of sight to the left. But I was getting edgy; the foot of the little rapid would be a good place to hold up and determine our exact position.

Both boats bobbed easily through the mild turbulence and we were quartering to pull for the left bank when another of the gray signs came into view:

DANGER

DISASTER FALLS

PORTAGE TRAIL 300 YARDS

and from downstream around the corner I could hear the low sound of a rapid, surprisingly unimpressive for Disaster Falls. It sounded wrong.

We scrambled up the high dirt bank and tied bow lines to a heavy log, then started downstream on a portage trail worn to dustiness by many feet. I couldn't wait to see Disaster Falls, yet I was afraid of what it might look like. Perhaps it would be so unlike any rapid I'd ever seen that I couldn't draw on past experience to cope with it. My mind was obsessed with its profile: if the slope shown on paper was full sized, and snow covered, people would be racing down it on skis! I tried to remind myself of the exaggerated scale of the profile sheet but that was little help and it seemed we walked the portage trail forever before we came to the head of the rough water.

Working our way off the trail and down along the bouldered left bank, we eventually found that Disaster Falls consists of six separate rapids, each running immediately into the one below it and each structured differently so that it posed a different series of problems. Collectively the Disaster Falls rapids embodied every obstacle, every maneuvering situation a fast-water oarsman would ever be likely to encounter. The sound had been unimpressive because of the distance from the head of the first rapid to the tail of the last: it plotted out at three-quarters of a mile, around three bends of river.

The map differentiated only between Upper Disaster Falls and Lower Disaster Falls and so we finally labeled the six "super-Upper," "upper-Upper," "lower-Upper," "upper-Lower," "middle-Lower," and "lower-Lower." Written, these names rather confuse the eye, but this helped us in talking about each of them and its relationship to the one just above, or below.

Super-Upper was a wide unobstructed chute with a symmetrical tongue of water pouring straight down between a few shoreline boulders on either side. Then the river widened a little and turned almost ninety degrees right. Upper-Upper seemed to be caused by an underwater mass of bedrock that narrowed the river severely and flumed it along the left bank between several immense, rounded boulders. Re-collected below this impediment, it immediately spread out three or four times its flumed width and some of it ran like a tumbling mountain streamlet left around a boulder-bound island; the rest ran straight ahead, wide and shallow, through a field of scattered red boulders we had named lower-Upper. Now, pushed back to the right by the bulge of the island, the next one, upper-Lower, gathered the water and charged down between the island and the right-hand shoreline cliff and piled half against an immense block of fallen cliff and half on past the block and over lesser blocks to create a thrashing, frightening hole. At this point the water coming around the left side of the island rejoined. Then for middle-Lower the river began a wide left turn through a field of larger but less numerous rocks than those above the island, and, to finish with lower-Lower, shallowed and made an S-turn, first through boulders and then against the right-hand cliff. Below that it rested; there was the first unbouldered, sandy beach in a mile.

We clambered over rocks and through an interminglement of standing pines and fallen timber for the length of the rapid — the six rapids — and had a first look at each of them. Some of the views were frustratingly distant: the island stood between the left bank and upper-Lower. We had to be satisfied with looking across the head of the island at upper-Lower's beginning, and up past the foot of the island to see it rush down the long chute and pile against the big rock.

My basic decisions could now be tentatively made:

Could each of the six rapids be run? Running each, would we emerge from it in the right location to enter the next one? Could we remember every hazard throughout the thousand-yard course and the specific point to start maneuvering for each?

On the way downstream I had mentally "photographed" each of the six rapids, and its relationship to the one directly above and the one directly below it. This gave me a kind of "double exposure" on each. At the end of lower-Lower it replayed like a film strip, but with parts still vague or missing. To fill in the missing images we retraced our route, walking upstream this time, stopping abreast of each rapid, restudying it, then moving back up to the next. By the time we'd returned to the trail near upper-Upper, my mind had all six spliced together as one long rapid, that in terms of structure, magnitude, and maneuverability of the boats, was runnable. The intangible ingredient, precision, was up to us as oarsmen.

Near upper-Upper we again intersected the winding portage trail where it came closest to the rapids and made the corollary decisions. We'd have to run without passengers; we'd need to be as light as possible for quick response of boat to oars and for reduced draft in the shallows. Both boats would run together rather than the *Norm* first making a "guinea pig" run; the course was long and intricate and if there was trouble, one of us might help the other, whereas no one from shore would likely be able to get to him. One boatman would help the other only if he, too, wouldn't be endangered or pulled out of position. No deliberate stop would be possible until the foot of middle-Lower, with any likelihood of being able to reenter the runnable course. If everything was okay after middle-Lower, we'd run on through lower-

Lower without trying to stop. All camera cases would be taken out of the boats and carried around, along shore.

The others headed down to pick out vantage points, while Fred and I went the opposite way to the boats. Yosh Oka, game to ride any rapid the river offered, had waited at the *Norm* in hopes of going through and I had to tell him he'd load me too heavily. He went down the trail especially disappointed, I knew, at missing the first of Lodore's big ones. It's as hard for a passenger to forgo an exciting ride as it is for a boatman, no matter how many stomach "butterflies" it may put to rest. It means he has done something less than the whole and leaves the rest until some uncertain future time.

Each of us busied himself with cockpit chores, checking safety snaps on canteens, unoccupied life jackets, and his spare oar. I was again running through my mental film strip of Disaster Falls to see whether all the images were still clear, and the sequence intact. When I finished my busywork and looked up, Fred was coiling his bow line. I crawled the steep dirt bank and untied from the log, then standing on the near part of the line to hold the boat, coiled the rest, dropped it into the cockpit, and jumped in after it. There was just enough calm water ahead for me to pull to the middle of the river and dip my hands over the sides to wet the oar handles.

The fast-water oarsman, like the bullfighter, has his own very private Moment of Truth. There is one at the head of each major rapid. It's not the point of no return, beyond which there's no going back to shore for another look. It's farther down than that, somewhere on the brink. It's a place, but it has to be described as a time. There's part of a second in which he can make one last judgment and one quick corrective stroke; there's another part of that second in which it's too late. Between the two lies his moment. A single word might summarize everything in the oarsman's mind that fraction of a second: "If . . ."

After that he becomes terribly objective, almost like a computer sorting its way through yes's and no's to the solution. He's picked specific features of the rapid as points to start each maneuver. The river carries him toward them. Start now? No . . . Now? Yes. His repertoire of maneuvers is elementary; he can drift, hold, pull left, or pull right. Left? Yes. More? Yes. More? No. Next problem.

Time is suspended for an oarsman in a rapid; there is only distance. Once, a few years ago, I remembered to check my watch somewhere near the brink of a rapid, and again below the tailwaves. It was a rapid of average length, perhaps a hundred and fifty yards long. Running it took forty seconds. I don't know how long we were in the six rapids of Disaster Falls; it seemed like more than five minutes and less than ten. I remember the narrow sluice of upper-Upper, and fighting, quartered right, to get started down a dogleg in the chute. I remember a sudden gust of wind that blew me out of position in the boulder shallows and tried to push me down the wrong side of the island. I remember wishing futilely for the momentary — just momentary — weight of a passenger on the stern at the foot of upper-Lower, and standing and throwing my own weight forward for whatever good it would do as the *Norm* plunged through and into the big thrashing hole. And I remember some of the rock-dodging and the appearance and location of some of the rocks, especially a blocky big red one that we had picked as a maneuvering marker in lower-Lower. Then I had the *Norm* beached on the sandy shore below it all and Fred was tied alongside with the *Camscott* and we had some bailing to do.

"How'd it go?" I asked. He'd been right behind me all the way from the top, but not having heard anything from him I hadn't risked looking back to see how he was doing.

"Great!" he answered, grinning. Then, reverting to his customary scientific understatement, "It seemed to go pretty well."

"Sure did. I'd like to run that one all over again! Thought for a couple of seconds that wind was going to push me down the left side of that island. Sure gusted up out of nowhere. Did it get you?"

"A little. But not like that. I touched a couple of rocks with my oars about there, though."

Then I remembered that I had, too. "Good thing we ran it without passengers." About that time I noticed my life jacket was soaked with icy water. It would have come from the waves in the hole at the foot of upper-Lower.

Each of the boats had two or three inches of water in the bottom of the cockpit. By the time we had dipped it down with bailing cans and were finishing up with the big sponges, the others were coming through the trees, down the end of the portage trail.

"Hey, that looked like fun," called Maggie.

"Most fun I've ever had in a rapid anywhere," I beamed. My arms and shoulders ached nicely with warm near fatigue; inside I was feeling the great jubilation a quarterback must feel after he has started near his own goal line and dodged and threaded his way through all the blockers and tacklers on the field for a touchdown.

"Beautiful!" exclaimed Doug generously. It must have been intensely frustrating for him to stand on shore and only watch, for he has "river" in his blood too, and is a fine boatman.

"If only I'd brought a third boat anyway," I thought, as I bailed, "after those last-minute cancellations." Disaster Falls and Hell's Half Mile have been so hard on rowboats, historically, that it had seemed an extra boat would mean extra concern. One of Powell's boats had been torn in half and then sundered to splinters on the rocks and shallows at the head of the island. The Kolb brothers had spent almost three grueling days lining their two boats down along shore and portaging equipment and supplies. Later voyagers had trouble there too. But for us, Disaster Falls hadn't turned out to be the problem I thought it might.

With the "D-Falls" behind us, we dug out the various jars and bags that comprised lunch, and ate it standing in the warm sun around the bows of the beached boats. For our drink, we even ventured to mix a bucket of Crash Orange. Then, with Triplet Falls and Hell's Half Mile waiting downstream, we loaded and moved on. Below Disaster, the Green resumed running as it had above the falls, narrow and hard-charging, dropping twenty to twenty-five feet per mile. There were more turns in it now; more maneuvering was needed to use the fast corners, yet hold away from shoreline rocks and ledges. Two more rapids, both fairly short and sharp, and then we had reached the head of Harp Rapid. We landed on the left to reconnoiter. The profile sheet showed an abrupt ten foot drop.

Jumping across the tops of big quartzite boulders, we made our way downstream around the corner and discovered the most photogenic rapid any of us had ever seen. Huge red close-set boulders lined both sides of the channel, like the nubs of worn old teeth, with the river coming around the bend and spearing down between them as a glistening tongue. The background cliff was in shade; the teeth and tongue of the river were sunlit. A flat rocky sandbar was abreast of the rapid and slightly downstream, and the extreme narrowness of the canyon put the whole scene near. As a final touch, the branches of a young tamarisk added just the foreground needed to give depth to the picture. If Hollywood built a rapid, I'm sure it would look like Harp Rapid.

With everyone else taking pictures, Fred and Doug ran the boats down in single file, off the long sleek tongue, and in to shore. Cameras were stowed in waterproof ammunition cans, and the run to Triplet Falls was resumed. Once, thinking we had arrived, I stopped above a noisy little riffle, but after checking and just getting started again, looked ahead to see another of the gray wooden signs. The red block lettering said:

DANGER
TRIPLET FALLS
PORTAGE TRAIL

Landing at the sign against a low silt bank covered with clover and jointed watergrass, we could hear and see Triplet's first drop.

Triplet Falls can be imagined most vividly as a huge "question mark," the "period" corresponding to the point where we had tied the boats. The river dropped down a straight little chute, then cut immediately to the right and rushed around the curve of the question mark, which was bordered by high cliff, pushing hard against it. Immense blocks of this outside cliff had broken loose, dropped down, and now tore a ragged edge on the water as it encountered them. Near the end of the half circle three chunks, larger still, finally pushed the river away from the cliff, mainly to the center and left, where it then encountered large boulders just out from the opposite shore.

Fred and I worked out our course and the maneuvers we would make at each critical point. On our way back up to the boats, we stopped for a few minutes to restudy the entry point and the first section of the rapid. This straight section of the "question mark" was rocky; we'd have to slide off the tongue near the bottom of it, and swing wide to the left to clear rocks at the lower center. Then as the river carried us around the corner we'd have to quarter severely and pull, to hold away from the cliff. There didn't appear to be any barely covered rocks or any waves that would preclude our running that section almost sideways, with the stern directly to the ragged water all around the outside of the turn. As we neared the big rocks at the bottom of the turn we'd ease up slightly on our holding strokes and let the water carry us just as close to them as we dared. Abreast the biggest one we'd pull opposites hard and fast to change from a left quarter to a quarter right; then we'd pull to hold away from the offshore rocks below on the left. When we'd cleared them, we'd land.

The run worked exactly as it was supposed to. We had had some reservations about the thrust of the water against the biggest right-hand rocks, and our ability to reverse quarters within a boatlength or so to avoid the lower left-hand rocks, but these proved to be only pre-rapid "jitters." It seemed the run was over a few seconds after it was started, and we were pulling in. With two of Lodore's three biggest under our belts, Hell's Half Mile just ahead, and a good beach at the foot of Triplet, we camped.

A thick cluster of box elder trees crowning the bar

behind us hinted at another interesting night. We had found that such jungles were the daytime shelter of the mosquito corps. They disliked sunshine or breeze, and stayed among the leaves and branches until the beginning of that soft, easy period just after sunset. Then they came out, brave and hungry.

Maggie unpacked her big rectangular mosquito net, and we watched enviously as she drove and tied the network of stakes and sticks and nylon cords that would hold it into box shape. She took no small amount of joshing about using it, and no one offered to help her with it. When she was finished, however, someone suggested it would be humane of her to let all the others sleep lying around the four sides with our heads inside the netting. She listened to the idea, made an expression that clearly said "nuts to you jokers!" and went down to the *Camscott* to make the salad.

Morning, like our previous morning in Lodore, was bright, but slightly cool until the sun began to warm the walls of the gorge. We had banked a damp upwind driftwood fire at bedtime and let the easy movement of evening air waft the smoke through our camp area. This seemed to diffuse the mosquitoes into smaller squadrons until it burned itself out sometime during the night. With morning, and the retreat of the shade line down the west wall and across our sandbar, most of them fled back into their cool, still lair in the trees.

Doug walked a hundred yards down along the left bank and reported that from there he could see around the corner. There was another gray sign within sight: HELL'S HALF MILE. We had been trying to reconcile the strip map of the river with the Dinosaur National Monument topographical map of the general area; each of these showed Triplet Falls and Harp Rapid in different locations relative to Pot Creek above and Hell's Half Mile below. My "dead reckoning," using approximate river speed and float time, led me to expect Harp sooner than it was actually reached. The same was true for Triplet. Thus my premature stop at the little rapid above Triplet — that rapid was where the topo map showed Triplet to be. The two maps differing, I had switched to the one showing the rapid farthest upstream. If that one was wrong, we'd still have some margin of distance before reaching the rapid's true location. Now, with the head of Hell's visible from the foot of Triplet, or just below, it appeared the strip map was right and the Dinosaur topo map wrong as to the locations of Harp and Triplet.

Landing at the head of Hell's, we tied to the trunks of stocky young box elder trees, and started down the portage trail. When it headed for gentler terrain away from the river, we left it and angled back toward the shoreline. Crossing a shallow wash, then working our way through a stand of small junipers, we made our way toward the loudest part of the roar, and came out on a rocky point overlooking Boulder Falls.

Most of the declivity of Hell's Half Mile is in the Boulder Falls section at the head of the rapid. Great blocks of red quartzite have rolled down a steep slope from the right and tumbled into the river, damming and confusing it. From shore they appear smaller than they are. With a boat among them for scale, most of the smaller ones are at least the size of the boat, and any single one of the largest would more than fill the cavity of a backyard swimming pool.

The river plunges against these. Then with most of its right side blocked by them, it takes a channel between an immense mid-river block and the left shore.

*Looking for a route through Boulder Falls. In Dellen-
baugh's* A Canyon Voyage *one of the Powell boats
was pictured just upstream from the rock where Doug,
Fred, and I are standing. Beyond the Powell boat a large
log was shown caught in the huge rocks. We believe the
log in this photograph, intruding from the right, is that
same log.*

Starting down that, reduced to a fifty-foot width, it piles into a left-bank gang of thirty-foot blocks that knock it back to the center, reducing its already reduced width by half, again. While the river is reeling from this one!-two! punch, it is also pouring down over a great spillway of buried rocks on its bed: Boulder Falls does all this to the river within sixty or seventy feet, then as a parting shot it splits the twice-narrowed river again with a great head of rock set in the middle of the falls near the bottom.

But that is only the beginning of Hell's Half Mile. Free of the falls, the river begins to spread out below as if overcompensating for having been confined and confused. Immediately it runs into two parallel boulder shoals that spread it far wider and thinner than it had intended, and divide it into three shallow, rock-studded chutes. These three eventually rejoin below, to make a whole river again. Of the half mile of Hell's, these shallows and shoals comprise all but the very abrupt section at the head that is Boulder Falls.

Our omnipresent three alternatives confronted us: run the rapid, line the boats through shoreline rocks along the edge, or take them completely out and portage overland. Of the three, the last seemed ridiculous, at least over the portage trail. Like the ones at Disaster and Triplet, it was picked and shoveled with canoers and kayakers in mind. The last quarter mile of it was gouged along the flank of the dirt talus a hundred feet or more above the river. Two men, even one man, might hoist a canoe or a kayak upside down and trek along the narrow path through the trees and high above the river. But not five men carrying a six-hundred-pound cataract boat sixteen feet long and five feet wide! Portaging over the tumbled rocks on shore would have been nearly as difficult.

The second alternative, lining, would be dangerous to both men and boats in the powerful turbulence of Boulder Falls, and practically endless in the shallows below. There was so little water along the edges of the chutes that the boats would have been out of it more than in, making it, too, essentially a portage most of the distance.

Lining or portaging could take a day, probably two days. Running would take a minute or two. There *had* to be a way to run it. We spent a long time on the big rocks overlooking the falls, and on the high downstream section of the portage trail, trying to put together an entry and a course through. We threw driftwood into the left-hand channel above the falls and watched it react to the different thrusts of the water. Where did the river carry a completely unsteered piece of float? How much could we modify that course with a pair of oars where necessary?

The big boulder "head" in the foot of the falls was barely covered. The darkness of it peered out like a face from behind a thin gray green mantilla of falling water. From what point did water starting down the falls go right of the big head, and beyond what point did it go left? Could we read those points from a boat?

After throwing sticks into the water for what must have seemed hours to the others, and discussing the river's treatment of each of them, Fred and I decided to try running it. We'd float down stern first, close by the great midstream rock, letting ourselves be carried back left, down to a point just above the gang of huge left-bank rocks. There a lateral wave angling in from a lesser rock would move us slightly right. We'd drop over the spillway into a lateral trough at its foot; that trough would almost stop our downstream motion. Then in the two or three seconds before the river could restore our momentum, we'd quarter left and pull, setting ourselves up to be carried down left of the great rocky head. Passing that we'd steer for the center chute of the three shallow ones. We'd "rock-dodge" at the top of it to get in, then run just left of center for a hundred yards down the chute, at the end of which there was a whole river again.

We were finalizing this plan when a half-dozen whiskered and hatted men in life jackets materialized from among the trees. We recognized two of them who'd been inflating the pontoon above the Gates of Lodore. Our group howdyed their group, and rather than start our run as we had been about to do, we stalled by prolonging our look at Boulder Falls. We hoped to watch them run it; watching a ten-man pontoon sluice through would be a better measure of the possible and the probable than watching pieces of driftwood.

Time passed: fifteen minutes, thirty minutes. Apparently they were waiting for us to do the same for them. More people began arriving, and then a thirty-three-foot pontoon with a half-dozen people on board came unexpectedly into sight, slithered down over the falls, and scruffed on through the shallows below.

Deciding Hell's Half Mile was getting too crowded, Fred and I started back upstream to make our run. The big pontoon had surprised us so that we couldn't remember which side of the big head it had been carried to. My impression was that it had slid up on it and then washed off to the left. Most of our sticks had been going that way.

On the trail halfway back to the boats we met Jack

Currey, a longtime friend, a "pontooner" whose operation and knowledge of river I respect highly. Jack and his wife Betty Ann were, they told me, making a Memorial Day run of Lodore with their children. We made a trailside ceremony of dividing a stick of gum three ways and started to talk "river." But then another pontoon pulled up and landed above Boulder Falls. We adjourned sooner than we would otherwise have, and each went to his boats.

Coiling the line and pushing off, I brought the point of the bow around upstream, and drifted stern first toward the head of the falls. Coming down to the midstream rock, I quartered right a bit and held, to let the current move me closer. Not until I was nearly at the brink did I realize the thrust from the left was materializing faster than it appeared to do from shore. I had entered at least a boatwidth too far right! As I slid over, I was still trying to regain part of it, even a foot of it, back to the left, but I dared not quarter more than a few degrees or I'd drop into the trough sideways and be rolled. All I could do was keep the stern downstream!

I was upon the great half-hidden head before I could grab a stroke beyond the trough. There was a sharp crunch, a momentary near stop with water roaring past on both sides, and then the heavy stern of the *Norm* had cantilevered out over the big dome far enough to pull the rest of it down into the sheltered water below. Luckily it happened so fast the boat didn't have time to swing sideways before dropping.

Fred, five or six boatlengths behind, saw all this, realized I had misread the thrust of the chute, and had a few seconds to make some corrective quartering strokes that entered him properly. From the calmer water below the falls I looked back to see the *Camscott* dive into the

*Early camp in Lodore, and a tin-cup toast to success
thus far*

trough, rise sharply like a surfacing submarine, and then pass easily and comfortably down the left-hand chute. We each zigzagged our way down through the boulder-littered shallows, and landed on a convenient beach. I envied him his run, with about two dozen people watching.

It was only midday, but we decided to camp where we'd landed. A Memorial Day traffic problem hadn't been anticipated in Lodore Canyon, of all places, and with "squatters' rights" on a good beach we decided to keep it. We had run all of the steep ten miles except about a mile, and the profile showed only two pronounced drops still below us. Hell's Half Mile was behind us, and I was anxious to check for damage to the *Norm.*

Breaking the routine was pleasant, even though it was only a few days old. We made lunch, then each of us did what he had to or what he wanted, reading, sunning, going over maps and notes, or just walking back up the portage trail to be alone. About midafternoon Doug, Fred, and I turned the *Norm* up on its side and located a small air blister where the big rock had impacted, separating the wooden hull and its fiberglass skin. It was easily fixed.

In the late afternoon the wind began rising, bringing low clouds with dark bottoms sweeping in from the southwest. By evening it was stronger still, and everyone slipped into a jacket or storm suit. Life jackets and other light items had to be snapped to the boats, and maps and hats weighted with rocks to keep them from blowing away. The driftwood fire blew horizontally rather than licking upward at the dinner pots and kettles.

In my duffel was a bottle of Courvoisier given to me

after a Grand Canyon expedition almost a year before. It was for some special time, or place, on this long run. There seemed none better than the foot of Hell's Half Mile, in a cold wind. Letting the cook-fire do what it would for the moment, we dug out the big white enameled tin cups. I poured a few dollops of brandy into each, and we bonked them together and toasted the wonderful Canyon of Lodore.

Five miles below Hell's Half Mile the Green escapes from the great north-south gash that is its Canyon of Lodore, and flows casually out into Echo Park. As it enters the north end of the park it is immediately joined from the east by the Yampa River, which has meandered across almost half the width of northern Colorado to meet it there. Echo Park is the riverine portion of a greater pocket around the confluence of the rivers known historically as Pat's Hole, and is in the Colorado portion of Dinosaur National Monument.

A late morning start from Hell's Half Mile, a long

stop at Rippling Brook Canyon to fill canteens from a feathery streamlet called Angel's Whisper, and a prolonged lunch brought us around the bend into Echo Park at 3:00 P.M. Looking ahead we could make out cars, and families at picnic tables along the high left bank. I pulled over against the island at the mouth of the Yampa, and beached the bow of the *Norm*. We seemed to have caught up with the previous day's traffic jam.

An appraisal of the island as a campsite, though, proved discouraging. The surface was still tacky from recent high water, and there were sharp cocklebur pods everywhere on the ground. Nothing was really stout enough or massive enough to tie to, and there was a nearly total absence of driftwood for cooking. More important, two feet of storm water would have covered the island, and there was always the chance of another flash flood down the Yampa.

Reluctantly, we drifted on downstream toward the people and the cars and stopped where a walkway had been stomped and scuffed slantwise up the twenty-foot dirt bank to make it climbable from the river. Climbing it, we found heavy posts had been set into the ground overlooking that portion of the bank. Good hitches dropped over the tops of two of these served to hold the boats against the moderate current.

Looking around after tying, we found we were at the edge of the grove, the south end of the National Park Service campground. An iron grate firepit and a plank picnic table were a few feet away. We considered continuing on downstream, but we were well ahead of schedule, and unsure of the existence of beaches in Whirlpool Canyon just below.

Our aversion was more to having neighbors than to having conveniences. We'd worked our way, a stroke at a time for five days, to get down the river to that point. Now here we were in the same campground with the cold-beer-and-fried-chicken crowd who had driven overland to it. We could hear youngsters and dogs farther up in the trees, but at least so far no one was camped close by. Perhaps *one* night, out of almost forty remaining ones . . .

Soup had just boiled when a lanky, sunburned man ambled into our camp. From the waist up, he was obviously a Park Service ranger, for he had on the fresh gray cotton shirt with arrowhead-shaped badge, and the disc-brimmed straw hat. From the waist down, it was harder to tell; he had on venerable old blue denims somewhere in that much-laundered stage between "just getting comfortable" and "about had it," and dusty, battered GI boots. If his hat had been a saddle hat rather than the straw, his weathered face would have been all cowpoke.

"Hello, folks," he offered, his thumbs hooked in his pants pockets.

I stepped out to meet him. He'd want to check our permit to boat through the monument. "Hello. Kinda crowded around here, isn't it?"

"Oh, Lord, no. They're just about gone now. You shoulda been here yesterday."

I was pretty sure we shouldn't have, but I knew what he meant. Memorial Day weekend. "Really bad then, eh?"

"Whooo-eee! I had 'em comin' out my ears! There was people comin' down the river, people comin' in the road, and people from — you name it. I don't know if I'll ever get this place cleaned up."

Apparently he was ranger, maintenance man, everything. "Well, I know we were piled up five or six outfits deep at Hell's."

"Yeah? How's your trip?"

"Just fine. Ran everything okay."

"I s'pose I oughta check your permit."

I started for the *Norm*. "I'll go get it."

"Aw, later," he said. "Whenever you get it unpacked. I guess I better get back to work. I've got to rake the area and move some tables — you know there's tables clear up there on that sidehill? You folks got plenty of firewood?"

"Oh, plenty. For tonight and tomorrow morning, too."

He started to amble off. "Okay, I'll see you later."

"That's a pretty casual ranger," offered Fred, when he'd gone.

"A guy like that sure makes up for a lot of the badge-happy 'seasonals' they get in some of these places," I suggested. "At least in my opinion."

"And gets just as much done," added Doug. "Maybe more."

A refreshing pool of late afternoon shade spread across Echo Park as the sun slid down behind the great Steamboat Rock monolith just across the river. The air stilled. A few mosquitoes droned around, but fewer than usual. We finished dinner and scoured and rinsed the pots and plates, put on a pot of tea, and were sitting quietly letting the day end.

"Well, I got a couple of them tables down off the sidehill!" It was the ranger again. "Beats me why anybody would haul 'em way up there to use 'em! But they sure did."

Fred offered him a cup of tea. "Oh, no thanks; I got some stuff on, over there at the cabin." He watched his foot scuff the ground.

"So you really had a big weekend?" Doug offered as a conversational handle.

"Oh, Lord! And it wasn't nothin' here. There was fifty-five thousand people up at Flaming Gorge."

"Fifty-five thousand?" My thoughts, and probably everyone's, flashed back to our two days around the reservoir just the previous weekend: a half-dozen forest campgrounds, large ones, each empty except for a camper or two. A whole day at Red Canyon Lodge with virtually no one interrupting our river shoptalk with the Reynoldses. Driving around to see Powell's Flaming Gorge and stopping to listen to an Apollo splashdown on the truck radio, along a highway that was nearly untraveled. The whole area had been practically deserted.

"Yeah, it was on the radio. All the stations up in there was broadcastin' that people should stay away because everything was full, campgrounds, overflow campgrounds, shoulders of the highways, there just wasn't anything left. . . . Well, guess my grub's ready." And then as if he were working for us: "Tomorrow I'll get the rest of those tables down, and get some of this cleaned up. See you folks later." And he turned and walked off into the grove of big trees, leaving us to finish our dinner and retire.

With morning, the sun and shadow effects reversed. Steamboat Rock had laid a long shadow across the river and into Echo Park the previous evening; now it received the morning shadow of the great mesa paralleling the river just to the east. During the hour or so the sun took to push the shadow down Steamboat's long yellow tan flank, each of us rolled out of his sleeping bag at some time or another and gravitated to the firepit. There were a few birds overhead in the large box elders and cottonwoods, but the campground was nicely free of human sounds.

While we were sipping coffee and letting the flames burn down to coals for a hot cake fire, we began to notice the ground squirrels scampering among the trees. They looked like chipmunks, but were larger, nearly as large as tree squirrels. At first they seemed to be just randomly hunting something for breakfast around the bases of the trees, but then apparently they smelled our cuisine and became interested in its fragrances. From some distance away, usually out of sight behind a tree or over the riverbank, two of them would advance tentatively, taking turns overlapping each other. One would come a few feet, stop, sit up, study us closely, and sniff the air. If all our movements remained carefully slow, the other would then advance from the rear and take up the frontal position. He'd sniff and scrutinize until he was satisfied; how one signaled the other to advance wasn't detectable, if there was a signal. And if the bravest was supposed to end up closest, this didn't seem to work out as it should either. When one had come as close as he could, he'd stand on hind legs motionless and unblinking for as long as he could bear it. Then it seemed panic would burst inside him and he'd break and turn tail, spooking his confrere just behind him, who joined him in racing out of sight amid a rattle of leaves and grass.

Another trip out, after two or three minutes of recuperation somewhere, brought the foremost one just a few feet closer; then, as before, he studied, he spooked, and they both sprinted. By then they were coming quite near Doug, and during one of their "mental rearmament" retreats he got out some cornflakes. When they began the next advance he was kneeling, holding one flake as far out to the little squirrels as he could, between thumb and forefinger. Seeing this from his closest point yet, the braver of the two made one heroic, dusty, leaf-rattling dash to Doug's hand, sniffed the cornflake, skidded like a batter rounding third, and was gone.

"Good thing you didn't flinch, Doug," someone said.

"Flinch! I just barely saw him!"

It seemed, though, that one flying sniff had told the squirrels cornflakes weren't what they wanted for breakfast, for after that they wouldn't repeat their raid for one. Doug tried tossing flakes out on the ground where they could be eaten without jeopardy, but they passed them by and came closer, interested in something else.

"Pancakes," theorized Doug. Taking one of Fred's "rejects" — the griddle had been too cold — he tore it and held out one of the pieces. Soon one of the ground squirrels was sniffing at it cautiously, then trying to snatch it to his mouth with little forefeet. When he held it firmly, to coax the squirrel to eat from his fingers, the animal retreated eight or ten feet and thought about it. Doug finally discovered that by holding two pieces of pancake between separate fingers of the same hand, he could induce the squirrel to nibble one while ambitiously eying the other. But when the first was gone, and the second was gone — zip! — so was the squirrel.

After many rounds of this they became more sure of us, and one eventually sat in Doug's cupped hand an inch or two off the ground, eating a bit of pancake from the other hand while we photographed him. We named him Fearless Fosdick, though "he" may have been a Fos-Jane, and a more cautious one Lop-ear, for its distinguishing physical feature.

The previous evening I had thought over our schedule and position, and at breakfast I told the others I wanted to hold the expedition over at Echo Park the rest of the day. We had made sixty-five miles in six days and had three more to make the remaining twenty-five to Dinosaur Monument. Whirlpool Canyon and Split Mountain Canyon were still ahead of us, but they held no rapids with the sharpness of Lodore's; twenty-five miles of such river could be run in a single day, if need be. It seemed useless to hurry past Echo Park only to hold over at some less beautiful place. It was quite peaceful there, after all.

About nine, the ranger came down, probably wondering whether we'd be leaving or not. But he didn't ask; he just scanned our camp to see whether we seemed to be packing up.

"I see the ground squirrels found you. They're pretty good little beggars. Golden mantled ground squirrel is what they are."

"Golden mantled. Yeah, we've been encouraging them a little bit," I admitted. "They don't care for cornflakes, but they sure go for pancakes."

"That one there's interested in somethin' else," he said matter-of-factly, looking over my shoulder. On the picnic table was a partial can of grapefruit sections. One of the squirrels had slipped by unnoticed, climbed the table, and was draped over the edge of the can, nibbling leftovers. On one of his frequent backings-out to see whether anyone was around, he saw us watching him. This seemed to give him pause, but only momentarily; then he ducked back in for another few bites. I eased over to my camera, got it, and took a picture of the grapefruit can with the last half of a ground squirrel hanging out of it.

"There's a couple of albino ones in here," he added, when I'd finished with the camera. "Have you seen them

*In Echo Park, looking north, back toward Lodore. The
Green enters the park at the base of Steamboat Rock,
left center; the Yampa from the right, just beyond the
grove of trees at the end of the road.*

yet? They just come out early and late because the light hurts their eyes."

I said that I hadn't, and changed the subject: "Major Powell's 1869 river crew were supposed to have carved their names on the side of Steamboat Rock or somewhere near. Do you know anything about that — where it might be?"

"No, I don't. But I don't think it's on Steamboat. If you ask me, it's back up the Green aways, maybe a mile or so. Now I don't say there is, but if there is, seems to me that's where it'd be. That writer James D. Horan had somebody out here looking for it one time but they didn't find it."

"You don't think it's on the slickrock above the junction of the river? Have you poked around up there?" I asked.

"Oh, yeah. I don't think so. See, the old cattle trail goes right through there, 'n I've walked that, clear up to Browns Park coupla times."

"From here to Browns Park?"

"Yeah, it goes up alongside of Lodore, up on top. See, that's how those old boys first got into this country. Old Pat Lynch, he come down here to Pat's Hole 'cause he found a note on his door sayin' he'd better get out of Browns Hole."

"You know quite a lot about this country," offered Fred.

"Oh, I teach school out here" — he jabbed a thumb toward somewhere — "in the winter. I been in these parts for about thirteen years. Do this in the summer. Trouble is, they expect me to sit around in dress uniform all the time and not do a damned thing. They had me in a desk job for a while, but they finally turned me out again."

"Well, you've got a pretty place to spend your summers," I said.

"Yeah, I like it," and then as if it had reminded him: "I got a couple of them sites raked up and burned. I guess maybe I'll get it cleaned up again, if I keep after it. I better get back to work. I'll see you folks later." He still hadn't asked to see our permit, or about our schedule, but I was sure he knew by then we had one, and would be staying over.

It passed pleasantly; there was absolutely nothing happening that we didn't make happen. It was limbo between Lodore and Whirlpool; even the landscape said so. The close-set blocky red walls had given way to walls of domed and rounded nankeen-colored sandstone, softer, lower, and set wider apart. Echo Park offered itself as gentle country.

At midmorning, Fred and Maggie went for a walk, which became a climb, to one of the points on the east from where they could photograph the long, narrow park. Sam and Doug disappeared in opposite directions and were gone through the middle of the day. Yosh studied his maps and notes and worked with his cameras. In the afternoon I walked to the mouth of the Yampa and then up along it for a time. We were like a family on vacation for a day, though most people seeing us float into the park would have held that the river trip was a vacation in itself. Still, the change of pace was important. There in that quiet little pocket in the Uinta Mountains I had a full day free from concern about high or low water, campsite selection, passenger compatibility, reprovisioning, boat repair, daily or overall scheduling. There were none of the problems that would be involved in the six hundred miles of river and countless rapids ahead.

Toward evening our ranger friend drove down in a pickup to report that he'd raked every campsite and burned all the burnable trash.

"You folks could use a little firewood," he surmised, looking at the few sticks we had left. "C'mon, I'll show you where there's some dead sagebrush. Makes a good hot fire."

Doug and I got up into the high cab of the four-wheel drive with him, and he drove to a flat at the downstream end of the park where large dead sage roots were plentiful among the live vegetation. We loaded the bed of the pickup, enough for our firepit and several others. On the way back, our benefactor lifted the shortwave radio microphone from its clip on the dash, and adjusted the tuning and squelch knobs slightly. "I guess I better call in and tell 'em I was just burning trash," he explained. "Somebody might've seen that smoke and got upset."

He stopped the pickup at a point along the road from which, judging by the care with which he picked it, radio communication through the rough surrounding country was best. "Headquarters, this is Echo. I thought you might have seen some smoke a while ago and just wanted to let you know I had a fire down here burning trash. It was pretty dirty, but it's all right now."

Back came Headquarters. "Echo, this is Headquarters. Please repeat that, you are coming in very weak and broken."

He repeated. There was a pause. "Relay, this is Headquarters. We just had a call from Echo and he's coming in weak and broken. Did you get any of it?"

"Headquarters, this is Relay," said a voice from some high mountain peak. "I didn't get much of it, but I think he said there was a fire down there."

"Relay, this is Headquarters. Do you see any smoke down that way?"

At this point our ranger went back on the air: "No, Headquarters, this is Echo. I said I had a fire a while ago but it was just to clean up trash and it's out now." But his transmission had been on top of Relay's answer to Headquarters, and garbled everything.

"Relay, this is Headquarters. What did you say? What did he say?"

"Headquarters, this is Relay. I think he said everything's okay, the fire's out now."

That seemed a wise point at which to leave it alone. Our friend sighed, put the microphone back in its clip, and drove us back to camp.

TO DINOSAUR

Day Eight began as Day Seven had, with clear sky,
retreating shadows, and birdsongs overhead. We had an
awareness of a bright, warm day but no time to luxuriate
in the details of it. There was packing to do, for it was
Back on the River Day: Whirlpool Canyon Day. Our
forty-some hours of layover had triggered real or imagined needs for nearly everything our two boats held, and
it had been brought up the steep bank a piece or two at
a time. Our camp had become a little solar system, with
the cooking area its center. Orbiting out at some distance in each direction was a bedroll, and around that

a satellite system of bags, pouches, and cases. All of these had to be rolled or tied or closed, and carried back down to the boats. There were the cooking pots and skillets and griddles, a daily chore. But the bedrolls, having been left unrolled for one day, seemed extra. Camera cases, maps, notebooks, and duffel — all had to be closed, carried down, and puzzle-fitted into the crannies they best fit. Canteens had to be carried across the campground and filled with water drawn from a hand pump. We six, collectively, probably made fifty trips down the steep dirt bank and back up, getting everything transported. By ten o'clock we were regathered and reloaded.

I had decided to have Doug row the *Norm* through Whirlpool Canyon. At Green River, Utah, many others would be joining the expedition, and I'd have a boat for him. Whirlpool, Split Mountain, Desolation, and Gray Canyons would hone him to a keen edge for the hard and fast rapids of Cataract and Grand Canyons, where we'd have to be sharp and precise. Some oarsmen of less experience would join at Green River, too, and I planned to bracket them, with more experienced men rowing the boat ahead of, and behind, each one. Doug was already a fine sophomore oarsman; I felt I could put him in number two position, run all day through anything runnable, and look back at the end of the day to find him still four boatlengths behind. Now the idea was to charge him temporarily with picking the best course for the *Norm,* and the *Camscott* behind it. He would, I was sure, notice new subtleties in the currents and new intricacies in the rapids. Sitting directly behind him, in the bow seat, I'd have his perspective and talk him through any course that he doubted.

Our first two miles took us around Steamboat Rock, which is a great peninsula of the yellowish Weber sandstone. For a mile we drifted along its high east flank, then turned its tip, then went nearly a mile back along its western flank. There the river swung away to the west and entered Whirlpool Canyon, plunging back into brownish red quartzite above which rose commanding cliffs of gray, blue-gray shales and limestones.

Doug handled the *Norm* competently, running continuously fast water and two pronounced rapids within the first hour. During that time I discovered what a compulsive thing "backseat driving" can be. It wasn't that he misjudged or mishandled; it was just that I didn't have the oars in my own hands. Sitting there behind him trying to learn how to teach, as it were, I found hitherto unrealized admiration for the self-control of a flying instructor who once sat behind me and rode through all my gross mishandlings of the airplane without sending me into further confusion by losing his own calm. I tried to emulate him and comment only when Doug waited longer to start a maneuver than I would have, but a lot of it was nit-picking. I missed any number of good chances to keep quiet. It's as difficult for me to be a passenger in my own boat as it probably is for some people to ride through fast fender-to-fender freeway traffic in their own car with someone else at the wheel. Not that they could do more behind the wheel themselves, but they'd feel better. Doug helped by being patient, and both teacher and pupil learned.

With the Yampa, one of its few undammed perennial tributaries, having come to join it, the Green was transfused; it became a strong river again. Above the confluence it had been a streamlet of waste water from a dam, seemingly emasculated, robbed of its strength, so that it resorted instead to the deviousness and intricacy of chutes and shallows and shoals. Below the confluence, though, where it was three-quarters Yampa, silty with spring runoff from the Colorado Rockies, the water had a new power and excitement. More than just the reactions of water to rocks and ledges that twisted and pushed at it, there was a fever, a vitality. Waves were higher and hit harder, like the flexing of the river's muscles. The current was more insistent, the roars of the rapids throatier, the slick tongues into them thrustier, and tailwaves rolled out farther, more sustained. I could sense the new river through the hull of the *Norm* as if feeling it through my skin. It was the nearest we would come to recapturing the essence of rivers before they were dammed. One of the group suggested that the river from the confluence down be renamed the Yampa, inasmuch as it was now the larger of the two, and the Green had been choked down to the function of a tributary to it. But with the end of the May-June runoff of snowmelt we knew the Yampa, too, would become a small river for nine or ten months.

By eleven-thirty we had left Colorado, floated a mile in Utah, and landed at Jones Hole Creek. Descriptions of this little stream had stirred my interest in it. One of Powell's men had caught twenty "fine" trout there in 1869; Galloway, in 1909, thirty-one on one day and ninety-eight the next. The Kolbs didn't try the fishing in 1911, but the Nevills party had in 1947, with good success. There was said to be a state fish hatchery a few miles up the creek, and if any fingerlings had escaped, they should have followed the water downstream toward the river, and perhaps grown to size there.

I mentioned the historical quality of Jones Hole Creek as a trout stream to Doug. "You know, you probably won't belive this," he said, "but I'm almost thirty and I've never caught a fish in my life."

"You mean you don't believe in it?"

"No — at least I don't disbelieve in it."

I asked him if he'd gone fishing very often.

"Well, not recently. In fact, I don't remember ever going fishing. But I must've gone fishing when I was a boy. But I don't remember ever catching a fish."

We decided to see whether we could reach a small milestone for him by having him catch one at Jones. Every man should have caught at least one fish, if only to know whether he believes in it or not. I probed into the bow compartment of the *Norm* and located a tin box of fishing gear I'd brought along, and a telescoping rod with spinning reel attached. We walked down toward the sound of the creek. En route we discovered that the big tree-covered bar was another of the boat camps, with tables and firepits, and many large, hungry flies.

Just downstream from the point where we intersected the creek it made a deep pool against a high bank, through some exposed roots of a big tree. I leaned out on the tree and scanned the pool, expecting to see the dark flitting shapes of many "fine" trout, but finally saw only one, about eight or nine inches long.

Rigging the monofilament line with two or three salmon eggs on a small snelled hook, and a small split shot a foot or so back for weight, I handed it to Doug and had him drop it into the upper end of the pool. From there the current carried it through. When the first two or three passes went strikeless I hiked along the creekbank for fifty yards to see whether there was a better-looking pool above. Not finding one, I returned to find Doug had a nine-inch trout on the hook. Not a monumental milestone, but a milestone. He was somewhat taken aback when I hit the fish on the head with a screwdriver handle, but I explained it made the hook removal safer, and was much more humane than letting the fish suffocate. He seemed to agree, but it later occurred to me that he might have intended to return the fish to the stream, and I wished I had waited to find out.

Doug rebaited and I walked the opposite way to look for pools between ours and the river. A few yards up the creek from the mouth someone had laid a driftwood and boulder dike across, presumably to keep trout from escaping out into the river. I checked the pool formed behind the dike and could see nothing in it, so returned to Doug's location.

"No use fishing the mouth of it," I told him. We thought there might be trout among the submerged tree roots at the foot of the cut bank on which we stood, but nothing more was caught, or seen. Discouragement, or perhaps reality, set in and we gave up. Fred returned from a climb up the high talus to sample the base of the Morgan formation, and Sam from a hike up Jones Creek. We again had lunch on a picnic table, wondering how long the campstops would continue to come "furnished."

That afternoon in the final four and a half miles of Whirlpool Canyon, we encountered at Mile 216½ a rapid more like the ones in Marble and Grand Canyons. At that milepoint the river is dropping over a hard limestone layer it has partially cut through, and that seemed to be the primary cause of the rapid. A chain of heavy ragged waves ran down the extreme right against the limestone cliff. There was a long, smooth tongue into the rapid at about center, a large hole caused by an underwater rock at its upper left edge, and then left of that a secondary tongue. Doug chose the secondary as I would have, put the stern down into the big waves, and we rode it out.

The whirlpools for which Major Powell named the

canyon are nothing more than eddies, though they were stronger there than anywhere above. The old, unfettered Green might have added enough volume and velocity to tighten them somewhat, but not sufficiently to endanger a boat. A person thrown from that boat, though, would be in some jeopardy. Probably because Powell's men rowed with their backs downstream, they inadvertently pulled into many of these and found the boats swung around by the perverse countercurrents.

Whirlpool Canyon became most beautiful just as it was ending for us. The left wall, up near the skyline, became an especially majestic battlement of crenels and merlons, with the afternoon sun on every promontory of yellow white sandstone, and deep shade in the thickly pined green coves between. Then just as the wall seemed to be building to a scenic crescendo it gave away to open country, and we turned right and drifted out into the open country of Island Park. There the river divided and fingered between several long narrow islands.

Thinking of campsites, I watched for a flank or point of island with beach, firewood, and suitable cover for the traditional "ladies upstream, men downstream" arrangement. The shorelines all seemed to be edged with a tangle of brush, which not only made landing and camping impractical, but suggested mosquitoes.

We continued. At one point my attention was caught by two things on opposite banks that seemed the complete antithesis of each other: on the left, a small nervous brown deer trying to be unseen against a brown sandstone cliff; on the right, a bare, lightning-killed cottonwood tree, stark as a pen sketch.

After four miles Island Park became Rainbow Park, and finally we found another "furnished" site. It had brushless shoreline, a table and firepit, and gnats as well as mosquitoes. But the view across Rainbow Park to the battlements of Whirlpool Canyon was splendid. And it was our last possible campsite before entering Split Mountain Canyon.

Doug had decided to complete his experience with the trout by having it for dinner, and so we got out the frying pan. Taking the fish from the bailing can of water that had kept it fresh, we rolled it in a flour and cornmeal mixture, and fried it. But he hadn't really enjoyed catching his first fish, and he was able to eat only half of it. He's not a predator at heart, and it will probably be a long, long time before he fishes again.

In the last two hours of daylight I walked up and sat on a long "hogback" ridge that ran to the river, separating Rainbow from Little Rainbow Park. To the west and north was the greatest sweep of open country we had yet been able to look into, a rolling, varicolored park of red and yellow and gray shales and clays, partly mantled by a sparse cover of sage and low juniper. Truly a topographical rainbow. On the crest of a ridge several miles north there appeared to be a small house and a high four-legged water tank. To the west, except for the shoulder of Split Mountain, I might have seen hints of Vernal, Utah, our first resupply point: mail coming and going, passenger exchange, laundry, groceries . . .

After a while Doug ambled up the ridge, hoping to find fewer gnats and mosquitoes there than at the river. We looked out over the country and talked, and by dark thought we'd solved most of the world's problems for it, which always seems so when we're on the river and the world is just a theoretical thing "out there" somewhere.

After breakfast next morning I unpacked a print of a photo taken somewhere near our campsite in 1871. The photographer, E. O. Beaman, had accompanied a second

Early morning, early June, Rainbow Park

Powell expedition two years after the first, and had
made a large number of glass plate negatives. Many of
these are still in the National Archives, and I had prints
of a number of them with me. If I could relocate the
points from which the ninety-eight-year-old photos were
taken, I wanted to rephotograph the same scenes in
color.

By holding the print of the Beaman photo in front of
me against the general scene it depicted, I hoped I could
walk to the point where all the ridges and skyline features were in the same relationship to each other as they
were in the photo. That point should be the original
camera station. My first try was back against the ridge
Doug and I had climbed the previous evening, then upstream in relation to the river. From there another long,
undulated ridge appeared to come into the scene as it
should have done, but there were no sharp upthrusts of
sandstone in the left foreground, or anywhere near.

Going downstream from camp, I discovered that the
ridge I was looking for was the one we had climbed.
Crossing over, then paralleling it, I saw sandstone out-
crops a half mile back and knew the foreground rock in
the photo must be one of them. After walking across the
sagebrush flat to them, I climbed two of the smaller ones
first, thinking that Beaman would not have carried his
bulky 1871 equipment to an extremely difficult place,
but was wrong about each of the two. Working up from
the bottom of the remaining and largest outcrop, I then
saw a point of rock that showed in the photograph. I was
seeing it in reverse; I had to be somewhere above and
behind.

It was difficult enough just climbing empty-handed.
I thought again of Beaman's suitcase-sized camera and
the heavy wooden tripod, and wondered why his photog-

rapher eyes had picked out that particular location to shoot from. My first try, nearly to the top, ended on a sandstone dome and too high. My second, down a cleft and out to a ledge, was too low and too far right. My third, along a nearly sheer face, out to an anvil point of sandstone, was closer. My fourth, back across the face and a little above, was within a few feet. From there, the point of sandstone that I had seen from below came into the left foreground. Little Rainbow Park rolled out beyond it, with the rainbow-colored ridge coming into the picture from the left. Over it, the great battlement of Whirlpool Canyon's south wall angled away into the haze. At my feet were two battered old junipers whose branch shapes revealed them to be the same ones shown in the ninety-eight-year-old photo. It showed them nearly the same size, then. If they had endured for a century, without changing markedly, how old had they been when Beaman saw them? How many other centuries had they lived?

Not expecting to range that far, I had come without my camera, and had to climb down and walk back for it. Fred offered to return to the camera station with me, and was able to refine my guesswork even further. Our retakes were within six inches of the original point and camera elevation Beaman had used. Undoubtedly he had gone to as much trouble in finding the point that suited him for composition as we went to in relocating it. The project took most of the morning, but recapturing Beaman's view across the parks was well worth the effort.

By eleven we had returned to camp, where the others were waiting to shove off and anxious about the rapids ahead. In its seven-mile length Split Mountain would drop the river another hundred and forty feet. Most of that would be at about twenty feet per mile, like the fast section that had exhilarated us in upper Lodore. But there were also four sharper drops on the profiles: rapids named Moonshine, S.O.B., Schoolboy, and Inglesby.

As we left camp, the two halves of the mountain loomed ahead like the Gates of Lodore had, with the river disappearing between them. Near this point one of Powell's men had observed that "the river seems to go for the highest points within the range of vision, disemboweling first one and striking for the next and serving it the same. . . ." A layman, impressed by what does seem to be so visually true there, might contend the river *had cut through* a mountain. A geologist would look at the pattern and grant only that the course of the river was *now* through the mountain, and he would know the river was there first, the land rose slowly entrapping it in its course, and then everything but the mountain eroded away.

A small rapid with a rock in the middle, just inside the entrance, served notice on us that the river was about to change from its slow, easy flow. Two or three hundred yards below, Moonshine Rapid roared up at us, impressively enough to merit reconnoitering. Landing on the left bank above it, we found that it had impressed others, too. A well-worn trail led downstream across the sandy sage flats.

The rapid was a long one, curving very slightly left and narrowing about midway down. At this point it was also roughest, with a wet-looking trough of holes running laterally across most of it like a waistband. Coming down to that was a long conventional rapid, of well-formed tongue running into tailwaves, and below it, a wider, rocky shallows.

"I think there might be an upset waiting for us in those holes," I told Fred and Doug. Looking it over, we decided we'd better "cheat" to the left and miss them.

Doug was still running the *Norm* and I asked him whether he had any qualms about trying it.

"No, if you think we can make it, I think we can." He did, with competence and precision, Fred close behind in the *Camscott*. Below, we found a course through the shallows without touching a rock.

A mile farther, we read S.O.B. Rapid from the boats as we approached it and ran without stopping, getting a good wetting from some great thrashing waves off our right. A mile beyond that we saw Schoolboy to be quite mild, and handled it the same way, then ran a long chain of racing water and landed on a small left-bank beach. We had run more than half of Split Mountain in much less than an hour, and I decided to call a halt for the day.

Early in the afternoon we saw a silver-painted pontoon coming down through the fast water just upstream, and recognized Mike Wintch, along with two other Park Service rangers from Dinosaur National Monument. I'd met Mike almost a year earlier, when our expedition was in the planning stage, and he had helped greatly in the logistical arrangements necessary at Dinosaur. Rowing the pontoon out of the fast water, he angled for shore and landed just above us.

We howdyed all around. "Say," he said, "we expected to catch up with you at Jones Hole."

"We got ahead of schedule," I explained. "Been having awfully good luck. Nice river." I wondered whether some emergency message for one of our group was his reason for looking for us. "We left Jones Hole about a day ago."

"Oh. I just wondered. We just came from there. Dropped the ranger off there for the summer." He went on to explain that they had put into the river at Echo Park, having driven in from Monument headquarters. "Did you do any fishing at Jones?"

I hedged. "We didn't have much luck."

"Just a minute, I want to show you something." He shifted some of the pontoon's cargo cases and brought out an ice chest. Pulling out a bundle wrapped in newspapers, he peeled them back to uncover a trout about two feet long. "German Brown. About three pounds."

Doug moved in and looked. "Where'd you get *that*?" he asked.

"Jones Hole Creek."

"Yeah, but *where*?"

"Right at the mouth of the creek."

Doug looked at it a moment and then looked at me and repeated my astute comment at Jones a day before: "No use fishing at the mouth." But he was smiling when he said it.

Mike rewrapped the big Brown and put it away, and we talked for a few minutes about Lodore and Whirlpool and mosquitoes, and the ranger we enjoyed so much at Echo Park. As he was about to cast off, Mike suggested meeting us at Split Mountain campground next morning and showing us a camping place we might prefer over it. We arranged to meet about ten o'clock.

Inglesby, the last of Split Mountain's named rapids, got us slightly wet the next morning; then within another mile we had emerged from the mountain and were at the campground beach. It was ten minutes after ten. By the time I had gathered up the things I'd need in town, Mike was there with a station wagon. "There's a picnic ground about three miles down the river that isn't used too often," he told us. "If you'd like, I'll drive you down for a look at it. If you don't like it, you can always stay right where you are."

We liked it. There was the omnipresent table and fire-pit, but the site was under a great, squat cottonwood tree about a hundred feet from the river and there was no one around. Fred and Doug said they would bring the boats down while I went in to Vernal to take care of reprovisioning and the countless other chores connected with the next ten days on the river.

I'd previously left a small motorcyle — a trail bike — in the Park Service vehicle compound. That was to be my transportation to and from Vernal, a distance of about twenty miles each way. To the carrier framework behind the seat I tied a briefcase stuffed with postcards everyone wanted mailed, expedition shopping lists, personal shopping lists for myself and each of the others, a list of phone calls to make, and a checklist of lists. On top of that I tied a plastic bag of laundry, and started for town.

Seven miles south, where the Monument's entrance road joins the highway in the village of Jensen, there was a service station with a short-order annex. After ten ice cream-less days, a stop there was top priority. It was so important I hadn't even put it on any of the lists; I knew I wouldn't forget. I suppose in today's living, that's "roughing it": ten days without a milk shake. Or a cold beer, or whatever one's favorite self-indulgence happens to be. The milk shake was so good I sat there on the bike and had a second one.

"Now this Vernal business," I thought to myself on the way in, "will be strictly time and motion. By early afternoon I should have it all squared away and be back in camp." As I blatted along on the motorbike I sequenced it in my mind. After coming into town I rode the length of Main Street, noting the locations of places I needed: sporting goods . . . post office . . . drugstore . . . pay telephones . . . laundromat . . . supermarket. Starting with the laundromat, I dropped the clothes into a washer, started it, and went to the post office. While the wash cycle ran, I mailed everyone's postcards, wrote and mailed my own, then went back and transferred the clothes to the dryer. Great. Now while they were drying I'd have time to buy some mosquito netting. The sporting goods store didn't have any. "No, I haven't tried the other place down the street." They didn't have any either. "Do you know who would?" They didn't. I was next door to a drugstore: the cigar and cigarette order. They had enough for half of it; if I'd wait they'd phone the distributor. "Yes, I'll wait." I looked through the racks for something for Fred to read; I didn't find anything he'd asked for. The distributor was out for coffee but he'd be right back. I tried the drugstore across and up a block. They had the cigars but not the cigarettes. "Yes, I'll take the cigars anyway." No book for Fred there either. They didn't know who'd have mosquito netting if the sporting goods store didn't. "No, I haven't tried the mercantile in the middle of the next block." Good, it was right across from the drugstore that didn't have the cigarettes. No, the distributor hadn't called back yet. "When he does, please cancel the cigars but not the cigarettes."

I spent the morning in that manner, finally finding Fred's books in an office supply store, and settling for nylon tulle, the stuff of bridal veils, in lieu of mosquito netting. Most of the afternoon was spent trying to locate things in a supermarket where I'd never shopped before, loading a half-dozen shopping baskets, and leaving a separate order of perishables to be added next morning. Before leaving town I made the phone calls I was supposed to make, and checked with the bus company to make sure they still expected to load and bring our provisions to us. Then I bought a cheap foam ice chest, loaded it with beer and pop and ice, strapped it on behind me, and rode back to Dinosaur. The sun was low enough behind me to cast my shadow far, far ahead as I putted into the campground where the others were waiting. And it wasn't until I unpacked that I realized I'd never gone back for the clothes in the laundromat dryer!

The cold drinks postponed dinner and turned it into supper. We had just finished when Mike Wintch stopped by to drive us to his campfire talk at the public campground. The Park Service had put together a slide show combining modern-day scenes with spoken excerpts from Powell's account of his expedition. They were showing this throughout the summer to campground visitors. The combination of slides with narration was very effective, and the voice used to represent Powell's seemed perfect as that of a robust, pragmatic, nineteenth-century scientist-explorer. When the slide program ended, Mike introduced our group to the rest of the audience and explained that we were retracing the still runnable portion of Powell's route by rowboat in connection with the centennial year observance. Then, to close the program, he got his guitar and played "Cool Water," dedicated to us.

At Dinosaur, we had completed the first of five segments of our expedition. Sam had arranged to leave us there, having previously run the rest of the Green by pontoon. The morning after the campfire talk he took the motorbike, promising to get the forgotten clothes from the laundromat en route, and started back to Flaming Gorge for his camper.

With little to do until our provisions and two new passengers arrived from Vernal, we remaining five decided to move the boats down to a closer point and walk up to the dinosaur quarry that is the reason for Dinosaur National Monument's existence. As we floated through the low ranch country below Split Mountain we could see the quarry visitors' center on a ridge to the north. The river's closest point to it was behind the Park Service residence area, and so we tied there and walked up across two or three hundred yards of sage flats, along one of the winding streets, and up the steep entrance road. From a parking area partway up the hill a rubber-tired shuttle train was hauling carloads of visitors on up the hill to the quarry. Apparently the monument had become popular enough to outgrow the parking area on top, necessitating another area partway up, and the string of little cars. Ours, like the others, had several bench seats running crosswise and a canopied top with plastic storm curtains rolled up to its edges. It reminded me of the lyrics about the surrey from *Oklahoma!* Fred would be most likely to remember: "Look Fred, it's got 'isinglass curtains you can roll right down.' "

He remembered. "Yeah," he grinned, " 'in case there's a change in the w-e-a-t-h-e-r.' "

The quarry building is built on the upturned edge of a rock layer known as the Morrison formation. The Morrison projects from the earth at a slant, and is "stepped" like the tops of different sized books leaning together. The building sits on a step that is lower, and is built against a higher one, having no wall there except the side of that higher step. Imbedded in this rock wall, and so enclosed in the building, are many, many dinosaur bones. Skilled workmen are carefully removing the rock matrix to expose many of the bones in bas-relief, a job that has been going on for years. It is slow work because the bones are relatively fragile, there are so many of them, and they are jumbled together like nails in a keg. Some stages of the work can be done with noisy, two-fisted air hammers, other stages only with dental picks and fine-bristled brushes.

We joined other tourists walking along the observation deck, an inside balcony that runs along beside the face of rock containing the bones. A recorded professional voice like that of the plane caller in an airline terminal bade us good morning, welcomed us to Dinosaur National Monument, and told us something about its history and what was being done there.

Twenty feet out beyond the observation deck were partially exposed dinosaur bones. Bones of all shapes and sizes. Whole bones and parts of bones. Leg bones, ribs, jawbones, vertebrae. Bones of little dinosaurs a few inches long and bones of monsters fifty feet long. Some shapes that looked like bones, and many, many that didn't. Signs posted along the railing had schematic drawings and captions to help us locate and identify important pieces. Diorama displays on another deck underneath had models and drawings of the many different kinds of dinosaurs, explained the climate and terrain during the time they lived, told which ones ate plants and which ones ate other dinosaurs, and how they all became extinct except the ancestors of today's alligators and crocodiles. Those ideas, the size of the great bones before our eyes, and the range of bone sizes from that of a man's finger to that of a man, were comprehensible. But not the sixty million years that have elapsed since dinosaurs lived.

We should have been rested and eager to leave Jensen the next morning. For two days, our feet had been dry, there'd been hardly any sand in our shoes, and we'd had relative freedom from the wear and wetness we call "boatseat-itis." We had a fresh menu of meat and bread and lettuce and eggs and such perishables. We'd had a change of scene and pace visiting the quarry. The call of the river should have been strong.

But it wasn't, this time. No one really felt rested. There had been the Vernal business for me, and nearly the same camp routine for the others as every day.

There were still meals to make, cooking gear to unload and reload, bedrolls to unroll and reroll. The boats had to be floated eighteen miles downstream from Split Mountain, eight of those miles after the visit to the quarry. They were slow miles. We found we had to row to make the pace bearable at all, for the river was dawdling and spiritless. Even the ranches on either side seemed to turn their backs to it; they fronted instead on the country roads. We could see the tall poplars and sycamores and painted posts along nicely fenced lanes. Those were the pretty faces the ranches would have presented from the other direction. For us, they were the backs of masks. Our view was of jerry-built pump houses of old tin and scrap lumber on crudely made concrete foundations, of stock corrals where cattle stood knee-deep in manure and watched us go by, of prostrated pigs sleeping in bogs unaware of our passing, and of old car bodies and bedsprings dribbled to the bank and thrown not quite in. That afternoon's ride past the unkempt ends of ranches reminded me of the indecorous behinds of stores that one sees from alleys, and of train rides through the frowzy sections of cities. The dreary uses of the river were depressing, and the low open country exposed them all to us. Its pace was equally depressing; there was nothing to remind us this was the same river that had charged so deviously through Lodore and pushed so strongly through Whirlpool.

Our arrival at Jensen might have lightened the mood, but the banks became increasingly tangled with brush as we approached, leaving no practical landing except a bare, smelly area directly under the highway bridge abutment where nothing whatsoever grew. Then, by climbing out from under the bridge, we alerted several hundred mosquitoes from the brush and they made a traveling swarm into town with us.

Thoughts of the river immediately ahead were no inspiration either. For the next hundred miles it would slog along through Ashley and Wonsits valleys at about the same gradient, one or two feet per mile. Until we reached Desolation Canyon there would be no rapids roaring up at us, nor even riffles to remind us of the river's capability. For me that was demoralizing; a raging river puts me more at ease than a silent one, and I would rather camp within sight and hearing of a rapid than beyond.

By walking along the highway for a half mile from the bridge we reached the little short-order shop that had detained me on the way to Vernal. About an hour remained until the van was due with groceries and our new passengers, Malcolm and Douglas McKenna. During that hour we were able to keep the milk shake machine mixing almost continuously. Then the McKennas arrived with Sam Carter following, he having retrieved his camper and driven back from Flaming Gorge Dam with the motorcycle loaded in back.

The only feasible place to deliver the groceries convenient to the boats was a piece of bare ground adjoining a government stock corral, across the river and a half mile down from the bridge. It also appeared to be the only unjungled or unbarnyarded camping possibility. Fred and Doug moved the boats down and the van was unloaded. Maggie systematically sorted everything into stacks, each kind of canned vegetable, canned fruit, bread, eggs, cookies, and we divided it into two piles, one for each boat. The mosquitoes were very pesty now, with the sun almost down and the air dead. We did the packing we could, had a quick dinner of corned beef hash, fruit and tea, and slid into sleeping bags theoretically out of their reach.

I had bought enough of the nylon tulle to cover my head and shoulders protruding from the top of the bag, but hadn't thought about anything to support it. With nothing to hold it away from me, the netting was no better than my shirt had been at the mouth of Swallow Canyon. No better for me; it *was* somewhat better for the mosquitoes because through the tulle they didn't have to stab blindly — they could see full well what they were jabbing their little needles into. I tried arranging it so its own stiffness and wrinkles would hold it away, but it insisted on collapsing, perhaps from the sheer weight of the horde of mosquitoes landed on it. The night was hot and I was sweltering in the bag, but it was my only real protection. When I last looked, Fred had abandoned his bag and put on a rubber storm suit with elastic cuffs and hood, and simply lain down on the ground. Shortly after that I hunched down into the bottom two-thirds of the bag and pulled the top and my bridal trappings down against me as a seal. While I was waiting to escape into sleep I could still hear the whines of a hundred optimistic mosquitoes; they were very audible through the hot, heavy bag.

Fred revealed his own frame of mind the next morning. When we had loaded and pushed out from shore he said he thought he'd row a while "to work off hostility." Clearly, the restless night and the languid river had put him in low spirits. Added to those was the sameness of the geology. Nothing dramatic seemed to be happening to the countryside. The river ran on the same strata most of the next hundred miles. There were no dramatic upthrusts or downwarps in sight to give fresh geological vistas every few miles. No great layered cliffs built by

Doug replaces the worn thole pin wrapping on one of our fifteen-pound oars.

advancing and retreating seas millions of years ago with the river's gorge cut through to reveal their stratigraphy. There were just the Ashley and Wonsits valleys and no contrasting landform asserting itself to divide one of the valleys from the other so that we could at least have a sense of transition. It was not an empty land, for there were dry farms, ranches, and oil and gas wells all through it. But it was empty of things to enthrall canyoneers. Two milepoints contained it like parentheses containing an incidental phrase: Mile 181, our Jensen camp, and Mile 70, the first definite rapid of Desolation Canyon.

We put our bows downstream and began to pull. Doug continued to row the *Norm*, even though the river offered no challenges against which to refine his fast water maneuvering. On the Wisconsin team, as there is on any rowing team, there had been a coxswain to set tempo and change of direction. Doug was accustomed to pulling in one direction until a change was called, and found it unnatural to look over his shoulder every dozen strokes or so to check his heading. The river made frequent checks and changes necessary by running through shallows and around low, bare islands of sand, as well as by its continually meandering course. The boats had an annoying tendency to jibe; they seemed to be temperamentally reminding us that they hadn't been designed to go down the river bow first. After a few oar pulls had them moving faster than the water, they would come about if the oarsman paused more than two strokes to check his bearing, then they would have to be brought around and pulled into motion again. The *Norm* frustrated Doug, reacting as it did so unlike a slim racing shell that lances truly through the water.

By noon we had made ten miles, hardly any of it due

to the current, which moved less than a mile an hour. About three miles an hour had been added by rowing. Fred, with Maggie, and Malcolm and Douglas McKenna had been ahead throughout the morning, intermittently in and out of sight around the bends. When we came upon them, they were beached on a bare sand island in mid-river. They had, Fred said, been making trial lunch landings on both banks for a mile or two, and at each the mosquitoes had driven them back to the boat. There being no shade and no vegetation on the slip of sand, it was nearly pest-free.

Partway up a gradual slope to the north an engine-driven drilling rig was laboring noisily at grinding a deep hole someone hoped would become an oil well. As we made our sandwiches and drank our flavor du jour from the bailing bucket, a big eight-wheeled tank truck rattled down to the river and backed up to a portable water pump. The boy who was driving looked out at us briefly, waved, and then started the pump. When the tank was full he shut off the pump and growled the heavily laden truck up the hill to the rig. There had been times, and would be again, when such an incident would have prompted bittersweet joking — that he had put at least half the Green River in his truck, or if he'd dumped in as much as he'd pumped out we'd have twice as much river to run on — but we were too dispirited to do more than watch, have lunch, and move on. Now only a hundred miles to fast water. By night perhaps only ninety. At twenty miles a day, perhaps only four and a half more days.

The tips of the oar blades made splashy spirals on the surface where Doug pulled them out at the end of each stroke. His strokes were long and efficient, from the waist and braced feet rather than just arms and shoulders. The boat moved, I judged, about ten feet from the point where a blade tip roiled the still water to the point where it roiled it again. Ten feet per stroke. Five hundred twenty-eight per mile, then. Fifty-two thousand eight hundred strokes to fast water, plus how many more for correcting jibes? How much energy in fifty thousand pulls? Either Fred the scientist or Doug the engineer could have computed it, but it seemed best not to mention anything that quantified our temporary disenchantment.

To prevent my impulsively saying something about the number of strokes still to be pulled, I turned my mental arithmetic on myself. Thirteen years, going on fourteen. Each of those years there'd been a thousand miles of river, perhaps more. Some of it sustained rowing like Doug and Fred were doing, but not all. Still, a lot of it. Perhaps only a thousand strokes per mile as an overall average. How many pulls of the oars in, say, twelve thousand miles of river? Twelve million. That many? Perhaps I should take only five hundred per mile as an average: that's only six million total. But what about all the late summer and fall trips when Glen Canyon was low and slow? And the rowing between the rapids since the dams decreased the flow? And rowing twenty or thirty miles of Lake Mead to get below the silt shelves so powerboats could tow us? I decided to take ten as a compromise; ten million oar pulls. Ten million oar pulls might make "cocktail talk" sometime. But they wouldn't have much meaning to anyone who hadn't pulled oars on the river.

Horseshoe Bend next belabored our impatience. As the name suggests, the Green makes an immense loop, first swinging east, then south, then west. At the downstream end of the bend it has taken nearly nine miles to return opposite the point of beginning, only a half mile farther down the country. The fact that we were measuring our progress in river miles, not airline miles, and that Horseshoe's nine could be subtracted from the hundred, no matter how deviously they ran, was only of intellectual comfort. Spiritually it was still frustrating to row all afternoon, dodging shallows and skirting islands, just to come around to the other side of a narrow neck of land we'd looked at four hours earlier.

Low, heavy, black-bottomed clouds had been moving in as we completed Horseshoe Bend and began to think of camping. At the turn where the river swung away to the left we were suddenly hit with a wall of upstream wind. Hard wind. Oarsmen swear the wind only blows upstream, and this was one of the winds from which such impressions are formed. It was so strong we could scarcely make headway against it. Doug and Fred yanked against the oars and everyone else crouched as low as possible to decrease the wind's purchase on us.

Our only hope within sight was the sandy upstream end of a long willow- and tamarisk-covered island a few hundred yards below the corner, and we made for that. As we closed with it, I could see the river was shallow at the point and down along either side of the island. We were above the right side so we took that, working slowly down abreast of it and finally making better headway after gaining the lee of some willows and tamarisks that were bowing and scraping to the pushy wind.

Jumping out with the bow line, I plunged shin-deep in silty muck and had to retrieve my shoes from the bottoms of the sinkholes my feet had made. Nearly the whole length of the bow lines was needed to reach across the bog and the head of the island to a large log on the far side. With the boats tied, we put on rain jackets and storm suits and waited for the rain.

Out on the river the wind was piling up frothy stacks of water everywhere, then seizing spindrift from those whitecaps and hurling it with such density that the height of the river seemed to be increased to the height of the wavecaps. All the fury of a typhoon was in the wind; only the shallowness of the river and its containment in its sandy bed, I am sure, limited the size of the waves, for the wind was belligerent and unrelenting. At sea it would certainly have been a full-blown typhoon.

Then after a time it changed. The wind tapered, stopped, reversed and began blowing back from the north, again insistently but much less so. Fred made a fire and we stood around it, still expecting rain.

Someone heard the sound first, then exclaimed "Oh, my God!" and we all turned and saw them, an immense cloud of mosquitoes. They were hanging in the wind a few feet above our heads and twenty feet back. There were so utterly many in the swarm that the mind has to resort to impressions: the feeling of the number in the swarm was like the feeling of the number of stars on a clear night. How many mosquitoes? Two thousand, ten thousand, a hundred thousand; none of these numbers feels like an exaggeration. So many of them that they stood out as a dark cloud several yards in diameter against the dark clouds of the sky. So many of them that we considered taking a picture of the swarm, knowing it would show plainly in a photograph. The whine-sound of the swarm was like a muted radio tone, or a sustained high from a violin string, and loud enough to come to us against twenty feet of wind.

They were such a compelling, unbelievable spectacle in terms of sheer number that it was impossible just to look at them and then casually turn away. Bits from old wartime newsreels came to mind, of waves and waves of bombers and fighter planes flying wingtip to wingtip above cloud, each machine surrounded by its own little buzz and all little buzzes being summated in the great reverberating drone of the armada.

Like thousands of tiny warplanes, the mosquitoes hung there just beyond us. As if the wind were barely strong enough to hold them back; as if they'd lunge forward and set upon us if it ebbed even the slightest amount. A million predatory eyes, half a million little stinging machines, all pointed to us. The battle line was drawn. Our weapons were mosquito repellent, a smoky fire, and the thickness of our sleeping bags. Their weapons were time, persistence, and having us outnumbered by many, many thousands to one.

Dinner was rudimentary as the previous night's had been: meat, fruit, and tea. When it was finished we dragged a great pile of driftwood, some dry and some damp, to an upwind location and lit it, hoping for a smoke drift through the sleeping area. It was still early but it seemed prudent to wrap ourselves in the armor of bed-rolls and plastic groundsheets before the stingered squadrons dove in. They were waiting; we could still hear them in the near dark, over the wind.

In the end the weather saved us from them. The wind had lowered the air temperature sharply, making a sleeping bag bearable so that one didn't unconsciously peel it open before morning for want of coolness or air. Rain fell intermittently during the night, the tap of individual drops almost countable on my plastic groundsheet. Each little shower served to rewet the driftwood that the fire dried as it burned, keeping a good smoke-screen going, and the wind held its direction fairly well so that the smoke kept streaming toward us. It was perfect anti-mosquito weather, the kind we had expected around Flaming Gorge and been surprised not to find except for the windy afternoon just below Hell's Half Mile.

Sometime before midnight a great roar suddenly burst through the wind and continued, sustained. It came from the east across the river and was the unmistakable sound of something blowing with indescribable pressure, like steam from many locomotives or exhaust from laboring jet engines. Malcolm and I were within talking distance of each other and guessed it to be escaping natural gas. We had come past several producing wells of the Horseshoe Bend Oil Field, with the big "horsehead" pumps nodding up and down in the distance. It was very likely someone had brought in a gas well in the middle of that stormy night. The roar continued for a half hour, perhaps an hour, pushed and pulled by wind gusts, but unabated in ferocity. I wondered how many millions of cubic feet of gas escaped before the roughnecks on the rig could cap a well like that one. Then, as instantly as the roar had started, it stopped.

The morning sky was still darkly overcast, but the rain held back while we had breakfast. The mosquitoes, sometime during the night had recalled the equivalent of several thousand fighter divisions, leaving only a few hundred patrolling the space over camp. The breakfast campfire smoke seemed to keep most of the remaining ones from using us as opportunity targets. In all, the night had been a good one — if restless, at least fairly biteless. It was a little unnerving to speculate what the situation would have been, had one stormy night — one out of twelve — not come just there.

We had made twenty-five miles from Jensen, five more

than projected. But at that new rate, even if we could average it per day, we still had three days and three nights to Desolation Canyon. The days on the river might be fairly mosquito-free, but what would the nights bring? Certainly not providential winds and rains each time we were ready to make camp.

A hard morning rain started, and fell while we loaded the boats. Then, as if to complete a perfect pattern that couldn't possibly repeat, a rim of blue sky appeared downriver and when we had rowed a mile or two, we were in sunshine.

Our target for the day was the village of Ouray, twenty-eight miles downriver. Between Jensen and Green River, Utah, a distance of one hundred eighty-two river miles, the Green touches no settlement except Ouray. Its fifty inhabitants are mostly members of the Northern Ute Indian Tribe, whose reservation lies to the south on the badlands of the Tavaputs Plateau. Ironically, like many other Indian reservations, the Northern Ute Reservation has been found incredibly mineral-rich, in this instance rich in oil shale, the equivalent of several million barrels of conventional petroleum. It was on the threshold of this region that John Sumner of Powell's 1869 expedition judged it to be "utterly worthless for anybody for any purpose whatever, unless it should be the artist in search of grandly wild scenery, or the geologist, as there is a great book open for him all the way." A rock formation from which oil can be boiled was unimagined in 1869. Shale refinery cities along the Green are hard to imagine now. But one more century is likely to bring them.

A late morning try for a rest stop proved that lunch on shore later would be out of the question. Our individual trips into brushy cover only stirred up personal clouds of mosquitoes for each of us, and they followed in hot pursuit as we sprinted back to the boats from all directions. At noon we tied the *Norm* and *Camscott* side by side so that each had one oar free for steerage, and made lunch on the decks while drifting. Our morning of rowing had yielded fifteen miles and the current grudgingly gave us a free half mile as we drifted during sandwiches, cookies, and punch. We had again improved our average.

At four o'clock the boats passed under the concrete bridge at Ouray and we landed just below. Our jostling of the young shoreline willows with the bow lines alerted the mosquitoes their branches hosted. Three Ute children swimming nude at the upstream side of the bridge seemed completely unbothered by them, but a swarm orbited around each of us as we scampered up the bank to the highway. Ouray was a fraction of a mile north; we could see an old white frame building with a soft drink sign hung from the gable and a telephone booth standing against the front. On arriving with our droning entourage of pests, we found that the white building was the trading post, but it was closed. Until then we hadn't remembered it was Sunday or realized it would matter. Across the highway were several small boxy buildings that were probably houses, but we could see no one watching us from the windows. There was no sign of life except the children at the river. The only thing that seemed to be open for business was the telephone booth, which had no door. Unable to get a cold drink or to get inside a building away from mosquitoes for a few minutes, or even talk to anyone, we took our mosquitoes and headed back to the boats, with Fred wondering out loud "how much they pay people to live here?" On the east side of the highway near where the children were swimming, a swollen dead horse with legs stiffly out lay on its side like a tipped-over toy.

There was little choice but to go on. The milepoints slowly came and passed: Mile 127 . . . Mile 126 . . . Mile 125 . . . Just above Mile 123 I thought a sloping shelf gouged from a clay bank looked usable and we landed there for a closer examination. It was rather narrow, but long, and I had decided to make it our campsite, when someone from the *Camscott* slightly upsteam called down to ask what I thought of it. Until then there had been virtually no mosquitoes — one or two — but when I answered "It looks pretty good down here," they were suddenly everywhere. It was as if little sonic sensing devices had locked onto my voice and zoomed the mosquitoes to its source.

They had homed in on the voices of the others, too. "Let's get outa here!" hollered Doug, heading for the tie in the *Norm*'s bow line.

Mile 122 . . . Mile 121 . . . Mile 120 . . . Just above Mile 119 a bald sandstone slope rose from the water to become a low gravel-covered hill. There were no tamarisks or willows on it, nor anything more than scraps of sagebrush for a hundred yards either upriver or down. On trying it, we were met somewhere near the shoreline by an unbelievably frenzied mosquito assault. Malcolm and I ran on up the gravel slope to see whether we could get above them. We couldn't. We skidded back down and bailed into the boats. The mosquitoes followed, swarming around the boats like bees around two oddly shaped white hives.

The next few minutes weren't very funny to us, though to an uninvolved observer they would have been pure Mack Sennett comedy. Fred was rowing the *Camscott* for all he was worth, trying to build up enough slipstream of air to leave his group's swarm behind. Doug was doing the same with the *Norm,* pulling as galley slaves must have pulled under the whip. This left the two of them at the mercy of a thousand sharp stingers, so the others had to protect their boatman as well as themselves, and at the same time try to whisk the hovering squadrons away into the slipstream. The scene for a mile or so was of two boatloads of people in total chaos, boatmen lunging desperately at the oars and everyone around them sitting, then standing, then turning, flailing the air, then flailing themselves, then flailing their boatman for a while, with maps, shirts, anything that flailed well. When it was over I had beaten the leather sweatband out of my hat and broken the brim.

Mile 116 . . . Mile 115 . . . Mile 114 . . . The sun was nearly down. Above Mile 112 we saw the late light catching something. Tents. And people moving around on shore. Approaching, we next saw their black rubber rafts tied to shore, and that they were on an island. Maybe a camp meant we'd run out of the mosquito country. A hundred yards closer and I could distinguish men, women, and children going through evening chores around tents and campfire. Then we had pulled close enough to see that each, no matter what he was doing, was pausing frequently to swat at himself or the air around him. It might have been practical to add our seven bodies to their camp, to spread the mosquitoes a little thinner among us all. But I doubt that even mosquitoes justify camp crashing; we exchanged waves of hands and took the short way around the bend, inside the island that divided the river there.

At five minutes before nine o'clock, with daylight all but gone, we beached on the head of a half-muddy island fortuitously laid up with two large snags of driftwood.

By the time we had carried the camp gear from the boats and sorted out enough good driftwood, darkness was complete. Supper was by the light of the cooking fire, in the smoke of two anti-mosquito fires, one at each upwind quarter. Our bedrolls were laid out on the bare upstream end of the island which was still wet from recent rain and too unyielding for scooping out contoured depressions for tired bodies. But spirits were good; twelve hours on the water had put us forty-five miles downstream from our last camp, and within forty miles of Jack Creek Rapid, where the river profile came alive again.

The pontoons we had passed caught us an hour out of camp the next morning. There were two of them, each powered by a small outboard motor, making a little more speed than we made rowing. As they drew closer behind and then began to come abreast we could study them, and they us. Each pontoon held a mound of camp gear, methodically stowed, and a family: father, mother, and teenage children. Their equipment suggested they'd done this kind of thing before; it wasn't spiffy and new, as if they'd bought it for the trip. The pontoons had been patched a few times but not painted recently, and the motors were several years old and of different makes. But the hats were the most valid clue. Neophytes would have had new hats; no one in either of the pontoons did. The women had touched up theirs with bright bands of new cloth, but the hats themselves, like the pith helmets of the men, were old and water-stained; they'd been down the river before.

The narrowness of the channel between shallows brought them close and they passed us, one right behind the other, within hailing distance. Their outboard engines droned evenly, stringing trails of bluish exhaust smoke low over the water behind them. We had rejected the idea of using motors before starting the expedition, and had been glad of it until Jensen. Then we had had some second thoughts about them; motors would have gotten us through the mosquito country faster. But now with perhaps only a day to go, things had begun to look better. Soon the *Norm* and *Camscott* could be turned around and run the way they were supposed to be. With rapids ahead, we were regaining our purist values. "What's that noise?" Doug asked teasingly of the people in the second pontoon, after we'd greeted each other.

"Oh, isn't that awful?" answered the woman, smiling pleasantly as they moved on by. Ahead of us, they read the river well, crossing and recrossing as necessary to stay in deep water. That removed any remaining doubt that they were river people, too, and made us less reluctant to get acquainted if we should encounter them further downstream.

Even back at the point where we passed their tent camp, the terrain had begun to show promise. Hills bordering the river had begun making steeper slopes, and occasionally the face of one broke out impetuously into a small cliff. They were increasingly higher, too, with every mile downstream. On the map the convergence of contour lines rather than the dearth of them changed the dominant tone from paper white to contour-line brown. Green tinting began to show away from the river bottoms, denoting upland forestation rather than lowland marsh. Place names of valleys and flats and bottoms began to be supplanted by place names of canyons, ridges, plateaus, and points. The river on the map was drawn narrower, with hardly any islands dividing it.

By Mile 100 there was the definite feeling of canyon. Brown sandstone walls rose vertically enough that they

A gradual upwarping of the strata, and the beginning of canyon again — the promise of rapids ahead

could truly be called cliffs, and high enough that they were our skyline and there was no view beyond them. The river was still slow, but fast water seemed attainable, and being able to look upward again at scenery and geology bolstered our spirits. At Mile 97, Major Powell's Bookshelves, a long row of immense sandstone blocks the color of old saddle leather, sat side by side by side by side on a flat hill below the skyline. Beginning at Mile 92 the western cliff took a sweeping arc around a great two-mile bend of the river. The massive wall was laid up of alternate layers of dark brown and light brown sandstones like an immense sandwich, with the arc of the curve so uniformly cut through them all that it looked as if one giant bite had been taken from it. From the base of this sandwich-layered cliff to the river, a slope of sliderock reposed. Protruding through that at intervals and set against the great curved wall were contrasting buttresses of darker sandstone that erosion had not yet been able to remove. One of our maps indicated the wall on the wrong side of the river and the others failed to show it at all. Powell's men had "given it" to one of their number, naming it Summer's Amphitheater.

At the mouth of Tobyago Canyon, a tributary at Mile 86.7, we encountered our first excited water since leaving Split Mountain. It was a small rapid, rating a doubtful one on the one-to-ten scale, but it was better than no rapid at all. As if that were our ration for the day, the river withheld any more riffles from us. An hour later, still rowing, we overtook the pontoon families camped in a brushy rincon of the right-shore cliff.

A mile farther, we found our own haven. A house-sized block of sandstone had fallen from a low cliff and out about twenty feet, leaving that space as a nearly hidden cleft between them. I had stopped to investigate as a formality; with the scarcity of sites we couldn't afford to pass any possibilities. There were nearly all the features of a good campsite, made even more appealing by being so unexpected. The level sandy floor, high sandstone walls on three sides, the mask of young box elders on the fourth, and the dead limbs strewn around for firewood made it almost ideal. It had the look and feeling of grottoes we're taught to associate with leprechauns and elves. There were very few mosquitoes, and a damp fire against the cliff filled the upper level of the little cranny with smoke that held them at bay. That night we cooked the steaks we had been trying to have since Jensen, and the rest of a full meal with them.

THE CAMPSITE GAME

Somewhere there *is* The Perfect Campsite — maybe. It has a sandy beach long enough for all the boats and is moderately sloping so that their bows slide up on it easily. There is no fast water pushing close by to complicate the landing or to chafe them against each other as they lie tied, nor any barely covered rocks to chew or gouge holes in them through the night. Within the length of their bow lines there is a large, well-rooted tree for tying each, or a heavy undercut boulder from which the line will not slip.

Only two or three steps up the beach is a dry, level

area that camp can be unloaded to, and within reaching distance of the campfire is a large snag of dry driftwood consisting of twigs for tinder, medium limbs for cooking, and heavy chunks for night logs. A clear little creek enters the river at this point. Back away from the beach is a ten- or twenty-acre domain of untracked sand, free of sagebrush and cactus, and gently duned to provide each camper with his own little pocket. At either side of the sandy section there are concealments for men's and women's "rooms." The site has shade when the afternoon is hot and sun when the morning is cool. There are no flies, mosquitoes, gnats, cattle, or other river parties. The view in any direction is a grand composite of Yosemite and Monument Valley. And of course this site is always just around the bend, whether one is camping early in the day, or running late.

Fred and Maggie and I have sought that river campsite for more than ten years. We've never found it, of course. And so we disagree as to how well any considered location measures up to our ideal. I've always claimed the campsite selection prerogative; a river expedition can't be run by committee, and decisions of where and when to end each day are concomitant with the setting of daily and overall pace. Yet there are usually latitudes in selection, especially on a small, unhurried trip, and we've been together on the river long enough to know what they are. Over the years we've developed, without ever intending to, a give-and-take pattern that I think of as the Campsite Game. It could even be formalized, and the rules handed out:

The Campsite Game

In the Campsite Game, the expedition leader is the Dealer and all others are Players. Any number can play, and all may drop into or out of the game at will except the Dealer, who is required to play every round. If some of the Players have been on a river trip before, playing can start almost immediately. But if they have not, the game can be improvised as the trip goes on, without instruction.

It is a "brinkmanship" game. The object is for the Players to make Dealer think he was out of his head in picking a particular time to camp, or place to camp, or, better yet, both. This is done by Bluffing. If the Bluff is not successful, the Players lose and have to camp where Dealer wants. That ends the game for the day. If the Bluff is successful, Dealer loses, and is penalized by having to look for an even better site, to start the next round. It will be seen that unlike the Players, Dealer can lose more than once per day.

The game is almost always played in the afternoon. Dealer begins it by deciding on a Campsite. The rules are that he must give first consideration to the boats, as they are the group's only practical means of transportation. Next, the Campsite must be good for the Players.

The Dealer's Handicap is that most beaches lack one or more features of The Ideal Campsite or, if they have them, the beaches will be encountered at the "wrong" time of day. (Dealer may choose an uncustomary time of day to camp, but this is vulnerable to Bluffing by the Players.)

The Players' Handicap is that the Deal may be handed over to them jointly or to any one of them if their Bluffing is careless or excessive. An analogy would be the old tradition, in hunting camps, of turning over the cooking duties to the first one who complained directly about the cooking, no matter how bad it might be. The words complained directly *are most important here. Direct com-*

plaints are too crude to be considered acceptable bluff-ing.

Bluffs are allowed in two general forms, Spoken and Unspoken. Spoken ones should be subtle and inferential. They are usually used by beginners and intermediates in the Campsite Game. Unspoken ones may take the form of either gestures or silence. They are usually the most eloquent and effective, and best used by advanced Play-ers. But there is no hard-and-fast rule, and a combina-tion of the two kinds also works well.

An example of a good Spoken Bluff would be one Player asking another loudly enough for Dealer to hear: "What do you suppose he's stopping here *for?" The word "here" can, with practice, be loaded with inde-scribable mock dread and derision. The burden is then on Dealer, for he must justify his chosen Campsite, with the added handicap that the Bluff has not even been ad-dressed to him.*

An example of a Silent Bluff, skillfully delivered, would be (after Dealer remarked he was thinking of camping for the day) the barely perceptible curling of a Player's upper lip, without speaking or otherwise reacting. Both of these examples are good because they don't quite render Players vulnerable to receiving The Deal.

The Campsite Game is at least one hundred years old. Take George Y. Bradley's journal entry for June 11, 1869, which we can assume was also expressed vocally within Major Powell's hearing: ". . . If I had a dog that would lie where my bed is made tonight I would kill him and burn his collar and swear I never owned him." New students of the Game would do well to study such classic Spoken Bluffs and to try emulating them.

But it must be remembered at all times that it is only

a diversion. There is never malice intended in the Camp-
site Game — any more than there is in the other games
humans play. Friendship and basic compatibility of
Dealer and Players is essential, so that all participants
know when to win and when to lose.

Doug and Malcolm had picked up the game quickly,
and in this, their second season of play, were doing well.
Our first game had been somewhere below Lodore, perhaps at Echo Park, but after that wind and rain and
long days of rowing had pre-empted several rounds.
With those behind us Fred, Maggie, Doug, Malcolm, and
I returned to the more regular routine and when not
preoccupied by the river, played the game down through
Desolation Canyon.

Compared to the land we had been coming through
for several days the canyon seemed majestic and the
name Desolation unkind. No longer were we out in mo-
notonous valleys whose silty floors were the dregs of
eroded mountains and mesas, but back among the mountains and mesas themselves; they rose up nearly from
the water's edge into walls and terraced cliffs and were
cut through frequently on either side by tributary canyons nearly as profound as that of the Green itself. The
word desolation suggests bleakness and something once
worthwhile now dead of futility and obscured by time,
but there was not that feeling. There was instead the
feeling of readiness; of incipiency, as if the land were
just about to happen. The great walls seemed new and
unworn. Slopes, so steep that only gravel would repose
on them, were planar and devoid of blocky clutter. At
first, trees bordered only the river. Not the slopes; that
would come in a later phase. Farther along, a solitary
pine was set into the right-bank talus halfway between

the river and the skyline. Then, as if that had been judged pleasing, a cluster of three or four more appeared several miles farther along; then trees on talus slopes became a regular appointment of the canyon. Nature's architecture in Desolation was very clean and solid and simple, drawn in lines either straight or agreeably curved. It seemed as though a beautiful visual reward was being prepared here for those who traveled the valleys above, but that we were seeing it before it was quite ready to be opened for official viewing.

Soon after leaving our camp behind the cleft rock, we ran a small riffle at the mouth of Little Rock House Canyon, and then flat water the rest of the day, until mid-afternoon, when we landed above Jack Creek.

First Player: "Are you stopping here?"

Dealer: "Uh-huh."

First Player: "Look over the rapid?"

Dealer: "No, I thought this would make a good camp."

First Player: "Oh."

Silence. Long silence.

Second Player: "What time do you have?"

Dealer: "About three o'clock."

Another silence.

Second Player: "We won't run it today, then?"

Dealer: "No, let's save it to start tomorrow. This is Mile 70 already and there's shade and good water."

First Player: "You want us to carry all the kitchen stuff down there beside the water and the shade?"

Silence. Everyone looks around. Players look at each other.

First Player: "Mmm."

Silence.

Second Player: "Do you need any help with that stuff?"

First Player: "No, I can get it, I guess. It'll just take time, and it looks like we've got plenty of that."

Dealer won that one and a few others, but he took a decisive loss later, at Rattlesnake Canyon, that wiped away any semblance of a winning streak.

At Jack Creek we were camped beside a noisy river for the first time in six nights. It was no profound bellow, but it would be enough to talk me to sleep. The little creek emerged between low cliffs two hundred yards back from the river, and rattled down to it over a bed of fist-sized pink and white stones. Our campfire and kitchen were on the north side of it a few feet from the clear water. Back away from the stream bed were several patches of sand that would be bedroll flats later. I had my eye on one near the river where I could listen to it, and could scoop out a depression to lie in rather than have to use an air mattress. There, after dinner, I worked on my journal, which had been neglected since Split Mountain and needed catching up before we challenged the forty or more rapids of Desolation and Gray Canyons.

On the opposite side of the creek, the downriver side, a fine grove of straight young cottonwood trees occupied a five- or six-acre flat running to the river's edge. High water from the Green and flashfloods down Jack Creek had merged just above the grove and swept through it, carrying away all underbrush and leaving a floor of flood-sorted stones in the rich shade beneath. It looked like a tree-clad plaza ready for grass and flowerbeds, and perhaps lampposts and iron park benches.

The two families of pontooners arrived toward the end of afternoon and tied just above the *Norm* and *Cam-*

scott. With the rapid clearly their prime interest they waved at us but walked as far out on the fan of Jack Creek rocks as they could, and looked it over. It was a long one, but mild and clean, with nice tailwaves. After a few minutes' study they returned to their pontoons upstream and when we saw them again they were shooting through, the glee of the youngsters' shouts rising over the rapid's low roar. When they'd cleared the tailwaves they landed somewhere at the downstream end of the cottonwood grove and walked back to talk.

We found that they were from Los Alamos, New Mexico, and had put in at Ouray, the day we first caught them batting mosquitoes on the island at Mile 112. For some reason Los Alamos, which is in the Pueblo Indian country, seems to have more river runners per capita than any community in the United States, all do-it-yourself-ers. Our new friends knew people I had met on the river and corresponded with about the river; it was one of those "small world" encounters. As we talked, their youngsters came running up with inflated air mattresses to ask whether they could run the rapid again. Before we finished our conversation about our trips and the river and the rapids and the mosquitoes, the youngsters had run Jack Creek Rapid several more times, carrying the mattresses up over the cobble rocks, then plunging through the waves lying prone and stroking with their arms for steerage.

Tailwaves of the rapid tossed a few of their splashy crests onto our stern decks next morning as we began our sixteenth day, whetting our appetites for the bigger ones that now lay just ahead. It was wonderful to be apprehensive again, to have to listen critically and look sharply and have so little time to make the right decisions before the Moment of Truth at the head of each rapid. Now all our oar strokes were once more against the current, holding to steal a few more seconds while we made sure, or across the current, moving laterally to set up an entry or miss a rock or a hole. Now a small midstream shoal of almost hidden boulders out from Cedar Ridge Canyon called for a fast choice between channels. Running one, we could then see that the other had been usable too, but even a choice point without penalty livened the river and sharpened our reflexes. Most split channels have a wrong side and it may, until the last few seconds, be out of sight below the brink or around an island. Now rocks from Flat Canyon shoaled the Green: which side was getting the most water? Did either side sound like angry, boat-breaking shallows? A choice made and run, then a hard right — almost a reversal of the river's direction — and another island dead ahead. Five seconds to choose. Down the right, but with some rock dodging. Now hard left of the river again and a surprise shoal on the outside of the bend: such things don't typically build on the outside. Then starting a fast downhill run of perhaps ten feet per mile through a chain of riffles and small, safe rapids. Now the river was exhilarating!

Our night camp, after some discussion, was made at the mouth of Steer Ridge Canyon. When the dinner plates and pans had been scoured, Doug and I slouched against a massive cottonwood log and somehow began talking about mosquitoes. Probably because there weren't any; they were far enough behind us so that we could be, finally, philosophical about them. What would be the difference in nature's balance, we wondered, if mosquitoes didn't exist? What if someone developed a sonic anti-mosquito machine that could either drive them away, or else lure them by the millions into some kind of death chamber? What if they were bred sterile and released, like fruit flies, so that their breeding eventually produced a doomed species incapable of reproducing new individuals? Would great numbers of small birds die — swallows, for example — and then the predators who fed on them, and so on, until ecology backlashed against man? It seemed the mosquito realm was far out of balance; there had been no greater number of birds in the sections of canyon where mosquitoes were so ultranumerous. I remember thinking as we rowed below Jensen that a man could almost certainly be brainwashed or driven to insanity if tied to a tree in the midst of a hundred thousand mosquitoes. Doug observed it was at least fortunate for man that something found in such profusion as mosquitoes wasn't as deadly as the rattlesnake, or there would be few, if any, of us still around.

Malcolm McKenna heard bits of our conversation, and came over to share our cottonwood log. Malcolm is a curator of paleontology with the American Museum of Natural History and has collected fossils from most countries of the world. He told us that he had at one time collected fossil mosquitoes from the Green River formation, a sedimentary rock layer that had formed Desolation's skyline for several miles. This provided an opportunity for the observation that "the only good mosquito is a fossil mosquito."

One of Malcolm's interests in Desolation Canyon was the possibility of encountering and sampling strata called the North Horn formation. He explained that it's a sort of paleontological bridge between the age of dinosaurs and the age of mammals. The North Horn was being deposited when the last of the dinosaurs died, and still being deposited when the first mammals evolved, and so contains the fossil bones of both ages. He had geologic maps, showing the approximate extent of each layer that was exposed by erosion and one of them showed a narrow east-pointing finger of the North Horn formation crossing the river, or rather, cut through by the river a few miles downstream. To determine where to start expecting it, we laid our topographic maps side by side with his geologic maps and compared the convolutions of the river on each. Comparing maps, we judged we might find the North Horn information about Mile 39, but for only two miles as the river crossed it. As far as Malcolm could determine, it had never been sampled in the Green River canyon, even though it should be handily exposed by erosion there, and we were anxiously anticipating the important things he might find. I promised him we would spend one of our four remaining days in the vicinity of Mile 39.

As the evening went on, Doug and I listened to Malcolm's intriguing accounts of fossil tiger bones being found in Alaska and camel bones in Colorado. We talked about the gradual but drastic changes in world climates that could permit such habitations and then terminate them. We talked about the evolution of man and the erosion of the land, about the pollution of air and water and about the chain reaction "greenhouse effect" that man has perpetrated on the atmosphere that sustains him. By then, the hour was late, the fire had burned low, and the night air was chilly enough for sleeping.

For two or three days, Yosh had been mentioning the Rock Creek Ranch, which he'd seen on a previous trip. "It's a living museum," he'd say. "I sure hope we have time to stop there." He made it sound intriguing and I'd checkmarked the printed map notation of it at Mile 54. Twenty minutes of rushing river whisked us the two and a half miles from Steer Ridge Canyon. Rock Creek, more nearly a modest little river itself, poured a fine charge of bright water into the Green.

Our view from the river was of an old brick building in a grove of tall poplar and mulberry trees, backed by a low juniper-covered dirt ridge. Between the boats and the old building was a long narrow field, one of the largest "bottoms" we'd seen in Desolation Canyon, paralleling the Green for a mile. It had once been irrigated; the long scratches of little water ditches were still evident, and scanty stitches of grass along the unditched strips between. Wires strung between old cedar tree posts held up the remnants of grapevines that crossed our end of the field, separating it from several rows of scraggly fruit trees that had, for a few years anyway, been pruned flat on top. We walked over a riverside fence of barbed wire that had fallen down, and followed the grapevine across the field to the buildings.

Only the walls of the house stood. There had been a fire and the roof had burned and collapsed onto the floors. The bricks we thought we had seen from the river were hand-cut blocks of reddish sandstone, of several sizes, laid up in a random pattern. The edges and angles were as true as if they had been machined, so true that no mortar had been needed and the joints between the blocks were only thin lines. The exterior face of each reddish block had been carefully dressed to a uniform rough-smoothness.

The main part of the house was a rectangle divided about equally between kitchen and bedroom-living room, with each room having one tall window facing toward the river and one other on the end wall. The door and window lintels and sills were great oblong blocks of stone, as carefully fitted as those of the walls. Off the

The old ranch house at Rock Creek. I had intended to pick up some memento of the place, but then felt wrongly about it, and went back to the boat empty-handed.

kitchen at the back was a small shed that seemed to have been added later as a tack room and pantry. It was not a large house, not the kind that could receive guests conveniently, but it had been more than just a bunkhouse or line camp for cowhands. It was a farmhouse, the kind a man would build that a woman would come and live in.

West of the house where the grapevine fence ended, was a log-cabin-like workshop. Several roof boards had fallen in, letting sunlight slash across the interior. As our eyes became accustomed to the darker inside, we could begin to make out known objects. Picks and shovels leaned against the raw log walls. Old wooden carpenters' levels and wrenches and hammers were strewn on the heavy plank workbench. Just inside the door to the right, a forge with homemade wood and leather bellows was built into the corner. Leather neck yokes and wooden doubletrees for horse-drawn implements hung from the roof joists, along with scraps of harness. Just outside the door lay a tangle of miscellaneous iron odds and ends that had been both raw material for the forge, and leavings from work done there. Looking over the mixture of things on the bench, Doug found a 1943 Montgomery Ward invoice for tools and equipment costing something over four hundred dollars.

All the horse-drawn machinery needed to turn, break, and plant the loamy soil was parked under the poplars and mulberrys. There were plows of several sizes near the house and shop. Out in the middle of the field with a clump of high grass shrouding it, a harrow stood alone with its pointed teeth still in the soil. It seemed the final statement of futility. Perhaps when the house burned the harrower stopped in the middle of his round and ran to do what he could; then, when it was over, didn't have the heart for anything except unhitching where it stood.

I had been going to pick up some memento of the place but then felt wrong about it and went back to the boat empty-handed. Everything there belonged to someone; perhaps some day someone or someone's son would want to ride back up the saddle trail, reroof the lovingly built house, hitch up the harrow, and finish the field. If one could live several lives, I think he might want to try one of them among the fields and peach trees at the mouth of Rock Creek.

Chandler was a moderately long rapid, rating perhaps a two or three, running against the right-hand cliff. Below it we caught frequent glimpses of the dirt road along the east bank, alternately gouged from the soft clay talus higher up, then coming lower and going out of sight behind the riverside screen of tamarisk and willow. We ran several savory riffles, a rather touchy shoaled section where the only safe water was along the left bank, and then a surprise rapid of extraordinary, booming turbulence. It was the first since Whirlpool Canyon's big rapid that had definitely been what I call a "life-jacket rapid."

By that time we were nervous that we would miss our two-mile chance at the North Horn formation. The gradually slanting upwarp of the terrain was carrying new strata up into view as we moved through it in a southerly direction. At Mile 40.5 Malcolm asked to stop to do some sampling of the Wasatch formation, after which we moved on, scrutinizing the shoreline strata for the gently rising sequence of layers to expose the North Horn. When it hadn't appeared by Mile 38 I called for a landing to reconnoiter. It was only midafternoon again, and as on the previous day there was not much interest in stopping. But if the geological maps were correct, we were already halfway through the two-mile exposure.

After we'd resolved that Malcolm should have one of the three remaining days to see what he could find around that milepoint, he got his rock pick and some sample bags and took a cattle trail down along the river toward where it veered against the left-bank cliff.

While he was gone, we unloaded the boats and carried all of the night's necessities up within the sagging arms of an old willow tree. During the remainder of the afternoon we read, and brought our journals up to date. Toward feeding time a mewly white-faced calf worked along the opposite shore, bawling for its mother. At dusk, with the late light filling it, the silty river became what Maggie described as "molten gold."

On the edge of darkness Malcolm came back up the trail, carrying a heavy chunk of gray black rock much too large for the pocket-sized sample bags. My hopes were high; what a treat to be witness to an important paleontological discovery. "Well, I've got something," he announced.

"North Horn?"

"No, Flagstaff Limestone. This stuff isn't even supposed to be in here." He went on to explain that an ancient sea now labeled Flagstaff Lake once occupied the region west of the Wasatch Plateau, the region now occupied by Salt Lake City and the Utah Valley. "As far as I know," he continued, "this has never been reported east of the Wasatch Plateau. So what I've done this afternoon, while you were all sitting around waiting on me, was to extend the known boundary of Flagstaff Lake about fifty miles farther east."

"Does it mean this formation was mismapped as the North Horn?" asked someone.

"What it may mean is that the North Horn is farther downstream than it's mapped. You see, this Flagstaff Limestone is supposed to occur in sequence between the bottom of the Wasatch and the top of the North Horn. Whoever drew this geological map apparently didn't know this other layer — the Flagstaff Limestone — came this far east and so they inferred the appearance of the North Horn too far upriver." He rolled the heavy chunk over on the ground. "This stuff," he said, pointing out fragments of fossil turtle and mollusk shells, "is just 'bone hash'; there's nothing important in it. But the presence of the strata here at this point is important because it seems to have gone unnoticed."

The next morning the others walked back upriver to explore another old ranch we had passed, and I became Malcolm's "bearer." Loaded with rock pick, canteens, geological and topographical maps, and cameras, we retraced his route down the cattle trail. After about a mile it pinched between the river and a tumbled slope of immense coffee-colored boulders. Leaving the trail, we hand-and-footed up over and around the great hunks for perhaps three hundred feet, to where there was a good overview of the strata across the river. At that point the Green and a deep tributary called Wire Fence Canyon converged, laying open a gaping cross section of terrain. Malcolm studied it all for some time, measured angles of upwarp with a pocket dipmeter, made notes and marks on his maps, and took overlapping photos with his Leica.

Then, picking our way back down to the cow path, we continued in the downstream direction, along a narrow shoulder trammeled from a clay bank above a noisy rapid, and across a rocky flat to the mouth of the next side canyon. A flat-topped promontory bounded the tributary, it being of the Flagstaff Limestone, cropping out

125

in horizontal layers. By climbing the slope we were able to sample the limestone at several levels, Malcolm knocking off preferred bits, dropping each into one of the drawstringed sample bags and labeling its tag. Pausing once for us to rest, he pointed out how ancient sediments, some from eroding red terrain, some from eroding white terrain, had intertongued to form the red and white banded deposit from which a spectacular cliff across the river had been carved.

At noon the boats picked us up, we ran the rapid Malcolm and I had climbed above, and crossed over to make camp on the west bank. After a hurried lunch, he took up his pick again and went out alone to investigate the strata that are exposed in Wire Fence Canyon.

In the middle of the afternoon our Los Alamos river friends landed across from us and we walked downstream a few hundred yards to watch them run Mile 36.5, sometimes called McPherson Rapid. When we returned to camp, Malcolm had also come back, with chunks of turtle and alligator bone, one whole clam, and a number of other fragments. Having made, in his career, far more intriguing discoveries than those little testimonies to subtropical jungle where there is now Utah desert, he was very casual about his afternoon's finds. On a recent trip to the Wasatch Plateau he had, he said, found fossil monkey bones.

"We can stay here another half day if you want to do more looking around," I offered Malcolm at breakfast next morning. But he was ready to move on, too. We hadn't found the mismapped North Horn formation, and the night had been unrestfully broken. There had been a rain that either prevented sleep or re-woke us and forced the hasty contrivance of cover and then, later, an invasion by a herd of range cattle. Apparently they wanted to come down to water, or to share the shelter of our cottonwood trees against the rain. Fred, who was farthest up the beach, said he'd been brought out of a sound sleep by the sudden awareness of a Presence and had sat up to find white-faced red Herefords all around him. He drove them away from everyone, but then only half slept the rest of the night, expecting to be stepped on. Our beach was wet from the rain and the morning sky was gray with somber clouds. The only cheerful aspects of camp were the breakfast fire, and the bright tarps and groundsheets of red, blue, and orange strewn around the yet unrolled sleeping bags.

At Mile 36.5 the river was leaving Desolation and entering Gray Canyon, the McPherson Rapid being its first of fifteen. Douglas McKenna and I climbed up and along a ledge overlooking the rapid to photograph Fred and Doug running it. The entry needed precision because the head of the channel was set with two large exposed rocks; then below them its right side ran furiously against blocks of fallen cliff. Fred went first, missing the nearest of the two rocks by a calculated oar's length, then quartering crisply to hold away from the right. Doug followed in the *Norm,* seemingly within an inch of Fred's course. The skill of the two men, the precision of their runs, was beautiful to watch from fifty feet above.

Five miles into Gray Canyon, the twelve foot drop through Range Creek Rapid was spread over sufficient distance to make it milder than the profile sheet suggested. But then, at Mile 26, Coal Creek Rapid reminded us that the river always saves surprises. It made an extremely angry mixture of rocks and water of even a lesser drop.

Approaching Coal Creek, we had already disregarded

Making beautifully exact entries around the rock at upper left, Fred and Doug bring Camscott *and* Norm *through McPherson Rapid, first of the Gray Canyon series.*

a narrow channel of water that split off from the right, and had passed it by, when at the brink we saw that it returned to the main body of the rapid and that it had been deep enough to divert a large volume of water around the head, leaving us shallows to float over in order to enter. The light into the cloudy water was reflecting from rocks on the bottom a few inches beneath the hull, and we moved over them with alarming speed. It would be marginal. I knew Doug was feeling what I was. "Stay with it," I cautioned, hoping I was right and wishing I was at the oars.

"I think I can pull back up to that channel," he said quickly.

"You're okay. Stay stern first. Work left a little." By then he was committed; our bad moment had passed, the bottom rocks were no longer visible and he had his work cut out for him. Everything dangerous roared by on our right as he held the stern into the ragged, swooping tailwaves. We rode them out and then turned to watch the *Camscott* alternately riding crests, then diving out of sight in troughs.

By midafternoon I was becoming anxious about saving some of the remaining mileage for the two days we still had scheduled. At Mile 23, camp seemed possible. An eddy in the lee of a rocky point provided an easy landing. The ground behind it was flat, but rock-strewn and brushy with clumps of sage and "Mormon Tea." On my map I'd already noted "no good campsites from Mile 37 to — ," leaving a blank to be filled later, and was ready to settle for Mile 23, where we stood. No one else had much enthusiasm, but Malcolm finally agreed it would do, if necessary. After a few more minutes trying to determine the true strength of each person's feelings,

I decided I'd better lose a round, and we went back to the boats.

A half mile farther, at the head of Rattlesnake Rapid, some bare sand showed through a sparse screen of willow whips, and I picked that as my next choice. It was a decided improvement; there was a little unwillowed pocket of level sand back against a low cliff. The cliff itself was a series of ledges, each more undercut near the ground, like inverted stair steps. With the sky heavily clouded again, the overhang seemed an important feature. I had the tract "sold" until Malcolm decided to take one last look, around the corner at the foot of the rapid.

When he came back up through the willows a few minutes later I knew I'd lost two in a row. We ran the rapid then, and landed at a fine sand beach the length of a football field, with plenty of driftwood, upstream and downstream concealments, and heavy logs to tie to. About all I had left was the opportunity, when it later started raining, to remind Malcolm that "my camp had nice sheltering ledges." And my ego suffered even more damage next morning when we passed three comparably spacious campsites within a mile or so.

That day, our twentieth, turned out to be our last above Green River, Utah. It was a Sunday, and within a few miles we began encountering bank fishermen who had driven up the riverside mining roads to try their luck with the catfish. Farther along there were families who had come in station wagons to cottonwood groves for picnics. As we ran down through Swasey Rapid, Gray Canyon's last, a dozen or more people looked up from sandwiches or watermelon to watch the little white Cataract Boats ride the waves.

That left only an impatience to finish. And the dam.

The dam had been a specter in the back of my mind all the way down the river. It was profiled on the sheets. Not a great drop, but nearly a sheer one, actually a man-made waterfall. I knew that pontoons sometimes slithered over it, but Cataract Boats draw more water, and don't "slither" well. The fact I'd never seen the dam made it even more of a bugaboo.

The dam is a diversion, to send water to a little power-house; the maps show it blocking the river at Mile 8.4. At Mile 9 I could begin to hear the sound, a low, even roar, not at all like the pulsing, ragged voice of a rapid. We tied where a gravel road neared the river bank. We could take the boats out on the road if we had to trailer them above the dam.

Scores of needle-nosed deerflies stung us as we walked the road to see the dam. We were still in river garb, cutoff jeans and swimming suits, and extremely vulnerable to their bites. Unlike those of mosquitoes, the bites hurt, and instantly so, and the flies didn't scare easily. They would land on a swinging arm or a walking leg with no hesitation and make their painful jab almost before we felt them on our skin. Our walk down the half mile of road must have looked like a traveling version of a slap dance, with one leg, then the other, lifted in the middle of steps so knees and calves could be batted.

The moaning monotone of the water drew us down a side road. We found the dam to be a bank-to-bank crescent of concrete, like a huge horseshoe curved upstream against the river. Its entire width poured over a brink — how deeply, it was impossible to tell — then slid as a fifteen- or twenty-foot-long sheet of water only a few inches thick down the steeply pitched concrete face into a pool of milling, disoriented water. On the far bank, perhaps three hundred feet distant, stood an old building that had been some kind of mill. Its huge wooden waterwheel was still in place, but hung inertly into the emptiness of a blocked sluiceway. On our side, just above the dam, a diversion canal to the powerhouse took a distressing amount of the river away.

It was fairly certain the water couldn't hurt us in itself, but the dam might. My main concern was whether I visualized the structure of the hidden concrete correctly from reading the behavior of the water that hid it. A backwave between the bottom of the dam and the pool of water just beyond it particularly bothered me. It was too smooth and uniform all the way around the curve. The boats should drop over the brink of the dam in any case, and skid down its face. Then if the continuous backwave were only caused by falling water colliding with dead water below, we would be all right — our momentum would carry us through it onto river that began moving downstream again. Suppose, though, that there was an upturned concrete "toe" cast into the dam there to give the falling water one last lift, up and away. If something like that were causing the wave, the boats would slide down the steep face into a hidden concrete barrier and there they would, in all likelihood, stay.

Only at two points was the backwave interrupted. One was at the opposite bank near the waterwheel, where an overflow chute was carrying a bridge of water down past the wave. But this chuting only increased the velocity of the water and then dumped it at the head of a rock-studded shallows.

The remaining possibility was a narrow tongue of water, almost the width of a boat, that bridged the backwave near the right bank where we stood, and shot onto the milling water beyond. The fact it existed gave me both anxiety and hope. It was the only exception to the continuity of the wave and suggested both that there must be some man-made reason for its uniformity, and at the same time that there, and there only, whatever was hidden by the water must be flawed, or broken away for a few feet. If the dam could be run in rowboats, I was sure that was the place to try.

Our return to the boats was essentially a gallop, with deerflies in tow all the way.

"Doug, I believe I'll run this one," I remarked, trying to be casual.

"It's all yours, buddy," he answered.

The barely moving water held back by the dam gave me almost too much time to think about the run. I was, in effect, betting that I had chosen the only safe point at which to fall over the dam, and that we could locate that point from the boats, at the brink. Second thoughts began to come, and we developed an alternate plan as we drifted closer. If I couldn't see the little spanning tongue from the brink, or if I saw something wrong we hadn't been able to see from shore, I'd row into the nearby beginning of the diversion canal instead of running the dam. Fred would follow with the *Camscott,* and we'd take the boats out of the river there.

Most of the way, everything was hidden by the river bend except the sound. Then we could see the bank-to-bank horizon of smooth water, then the river downstream, then some of the roiling pool below the dam. I tested the thrust of the current and found it was possible to hold against it, and even to make some headway back upstream. All of the pool showed. All of the long backwave showed, with the little tongue slightly to our right. Everything segued into slow motion. Hold. Look. No new dangers showing. Plenty of time. Quarter and move right, to fall directly above the break in the wave. Quarter back, get the stern right-angled to the drop. The transom was at the brink, hanging over. Now! I pushed both oars hard to make sure we carried over. The *Norm* pitched — down abruptly. There was a braking of our momentum and the scuffing sound of fiberglassed hull on wet concrete. It could only have been two or three seconds, but we seemed hung on the face of the dam for minutes and I remember thinking we were going to stop sliding, and hang there like a ridiculous ornament. But we slid and the stern found the break in the backwave and plowed through and we rose onto the apron of water beyond. We turned in time to watch Fred's boat pause on the lip, fall away, take its own slow-motion slide, and shoot up through the wave into the pool behind us.

There was no enthusiasm for the Campsite Game on the afternoon of the twentieth day. Dark, foreboding clouds had moved in again, as they had been doing every afternoon lately. We were halfway down through Utah, and it seemed the farther south we rowed, the more un-settled the weather became. Camping at midafternoon now had no appeal for any one of us, with less than eight miles remaining to the town of Green River. A phone call would bring our cars and trailers to meet us one day early.

With the near certainty of rainstorms behind and the likelihood of milk shakes ahead, we turned the bows downstream and pulled for town. When we began seeing rows of junked car bodies riprapping the banks of the river, we knew it wasn't much farther.

CRY MY BELOVED CANYON

The story of our six days between Green River, Utah, and the head of Cataract Canyon could be told as a continued account of rowing on slow water, playing the Campsite Game, and slapping mosquitoes now and then. But there is, I think, a better story and a more important one for me to tell at this point. It's the story of a seven-year love affair that began on the river, and that I took for granted until it was ended. It's written in sorrow, about my splendid, lost love, and how part of the Green River Canyon almost restores some of it to me. This story is about Glen Canyon.

The creation of Lake Powell had been ordained by the time I discovered Glen Canyon for myself, and its superlative loveliness now lies cold and mucky under the darkness of the lake behind Glen Canyon Dam.

Even though I knew its days were numbered, I took the canyon for granted, as one does familiar persons and places. When the first wonderment had been diminished by repeated passenger-carrying runs of my guide service, the Glen became something different, something less. I forgot its true vitality and worth. Disenchantment came easily on the low water of late summer, after having run the tumultuous rapids of the Grand Canyon. Then, the Glen was anticlimactic. I rowed and floated it many times in the seven years the dam was being built, and other times used outboard motors or drove high-powered jetboats. I was often impatient with the slowness of the current and disgusted with the shallows that gouged and scraped my boats. My fondest wish then was for deeper water. Three or four more inches would usually have been enough. Instead, when it came behind the dam, it was three or four hundred feet.

Given that wonderful canyon again, I might sometimes still have the same feelings about low, slow water. But now I would know, really know, what the alternative is. The filling of a canyon is abstract until one sees it happen. Then it's reduced to realities that hurt: friendly shoreline suddenly gone forever, curtains of dead water half ascended up dramatic tapestry cliffs, favorite grottoes glutted with flotsam from inundated beaches that had been campsites. The highest branches of venerable trees reaching up out of the water like the desperate hands of the drowning.

"Isn't the lake beautiful?" is the question asked all around it. The answer is yes. It is beautiful, despite the ugly death that gave birth to it. But that's not the question that's really being asked. By tone and implication there's always another question between the lines, one that rather anxiously seeks reassurance. Sometimes someone who isn't careful will even ask it directly: "Isn't this prettier now than when it was river?"

It isn't necessary to be making money from Lake Powell, or have hopes of doing so, to pronounce it beautiful. It's more difficult to be avid about it, though, if one has known both the lake and the incomparable canyon it covers. Then I think one is entitled to claim a more objective evaluation.

But it gains nothing to discuss relative beauty now; beauty never had a chance. Perhaps it will some day. In the meantime, the prevailing twentieth-century values said: Glen Canyon is *useless* until we fill it with water, no matter how much beauty we cover.

Major Powell wasn't writing about Labyrinth and Stillwater Canyons on the Green, but could well have been when he wrote these sentences about Glen Canyon: "The smooth naked rock stretches out on either side of the river for many miles. . . . Sometimes the canyon walls are vertical to the top; sometimes they are vertical below and have a mound-covered slope above; in other places the slope, with its mounds, comes down to the water's edge. . . . One could almost imagine that the walls had been carved with a purpose, to represent giant architectural forms . . . curious, narrow glens are found. Through these we climb by a rough stairway perhaps several hundred feet to where a spring bursts out from under an overhanging cliff and where cottonwoods and willows stand, while along the curve of the brooklet oaks grow, and other rich vegetation is seen. . . . So we have a curious ensemble of wonderful features — carved walls, royal arches, glens, alcove gulches, mounds and monuments. . . . Past these towering monuments, past these oak-set glens, past these fern-decked alcoves, past these mural curves, we glide hour after hour, stopping now and then as our attention is arrested by some new wonder . . . we walk to explore an alcove which we see from the river. On entering we find a little grove of box elder and cottonwood trees, and turning to the right, we find ourselves in a vast chamber carved out of rock. At the upper end there is a clear deep pool, bordered with verdure. . . ."

This was Glen Canyon in 1869 and until 1963. Like nearly everyone who knew it well, I thought of it as mine. I'd sung in Music Temple, resolved the riddle of Mystery Canyon, climbed to explore above the choke rock in the Hidden Passage narrows, waded a flashflood to beat darkness back from Rainbow Bridge, had some near falls from precarious "moqui steps," bellied down for cool drinks from dozens of the canyon's spring-fed streamlets, and scooped out sleeping bag cavities on nearly all of its good sandbars.

The Glen was a wonderful do-it-yourself canyon, and the Colorado through it a safe river even for first-timers. Anyone could run it and did, in nearly anything that would float. Canoes, kayaks, all kinds of small boats, huge and tiny pontoons, truck tire inner tubes, inflated air mattresses, life jackets. It was even swum once, in rubber wet suits. There were no rapids, only a dozen mild riffles, and they could either be navigated or bypassed along shore. There were fresh water springs for refilling canteens, sandbars and rock ledges and the mouths of canyons for campsites. It could be floated through without power in seven to ten days, or motored through hurriedly for the "scenic trophy" tourist in

Labyrinth Canyon on the Green: fleeting glimpses of "The Glen," now buried beneath Lake Powell.

three or four days. Roads and even airstrips were close by at either end, making logistics quite easy. It was a warm, quiet, intimate little world of its own, a fraction of a mile wide and a hundred and fifty miles long. It was mine for seven short years, while Glen Canyon Dam was being built.

Labyrinth and Stillwater Canyons on the Green are the nearest I will probably come to refinding what was taken. Running them for six days was a bittersweet experience because they are so remindful of the Glen. It's as if that unsurpassable canyon had been disassembled down to its smallest component features, then those jumbled and put together again, not quite right and with some missing. Many of the great mounds and cliffs wanted to fit themselves into my mental filmstrip of the Glen. Mouths of new canyons hinted that old familiar Indian ruins should be found inside. There were beaches I was sure I'd camped on, and tapestries of desert varnish on brown sandstone walls that counterfeited the originals. Like Glen Canyon, the Green, here, was gentle, and shallow enough so that the youngsters in our new group could float alongside in life jackets, or swim, or run along on the sandy bottom in knee-deep river. On several bends, rocks that I must have maneuvered past before, too, protruded above the surface.

For a hundred miles or more, the sides of my eyes caught fleeting glimpses of my beloved lost canyon, but when I turned to look squarely it was never really there. It was the same sweet anguish as having only droplets of water, when one longs for a deep, refreshing drink.

TWO

THE COLORADO

CONFLUENCE

The Green is born in Wyoming's Wind River Mountains, the Colorado in Colorado's Rockies, both craggy, basin-bordering ranges punctuated by thirteen-thousand- and fourteen-thousand-foot peaks. But they can't be called rivers until later, when they have run several hundred miles and cascaded downward several thousand feet in elevation. In the mountains there is nothing to disclose to whom the infant waters belong or what they will become. Snow simply melts; trickles of water converge; urgent rivulets begin to find each other; streamlets soon come by the hundreds to form a young Green and a young Colorado hundreds of miles from each other, tumbling down through high country secretly gathering their tributaries. When they have gathered enough of them to be noticed, man filches from them. The waters that have thus far survived his diversions and diminishments run out across the desert. They meet at a Y-shaped confluence, in a thousand-foot gorge, in the cynosure of Utah's awesome Canyonlands National Park. From there, the conjoined rivers are the Colorado River, running in a deep trough of rock across the American Southwest to the ocean; the Sea of Cortez.

Down the Green to the confluence of the rivers, we had brought a confluence of people, professions, and personalities: an attorney, a banker, a musicologist, a paleontologist, a teacher, a lumberman, and businessmen, housewives, and children. Passengers from school age to retirement age, from seven different states, rowed down the river by oarsmen from four.

Our three-hundred-mile run from Flaming Gorge with only two boats, two passengers — and they river-wise — and the thrill of new canyon, new river every mile, had been a boatman's holiday for Maggie, Fred, Doug, and me. The addition of five more boats and sixteen more people ended our sabbatical, but without regret, for we found in the ensuing days that we had assembled one of the most enjoyable river parties imaginable, one whose humor and adaptability would carry us through some bad moments, bad hours. An unpleasant river passenger is called a "clinker"; there would prove to be none among this group.

At Green River, Utah, we had disbanded for three weeks, having run in twenty days a length of river that had taken the Powell party forty. The interlude had been scheduled for several reasons: to take advantage of better boating water through Cataract Canyon, to allow time for reprovisioning and for boat repair if necessary, to allow ourselves a physical and emotional rest at approximately the midpoint of the seven-hundred-mile expedition, and to depart Green River on July 13, the same day as the Powell departure one hundred years earlier.

Returning to Green River after the interlude, we were joined by John and Mary Womer of Chicago; George and Charlie Philips of Lansdowne, Pennsylvania; Rukin and Keri Jelks and their three sons from Phoenix; Win Harper from Cathlamet, Washington; Malcolm's wife Priscilla McKenna from Englewood, New Jersey; and my wife Joan from Flagstaff, Arizona. To the *Norm* and *Camscott* we added *Mexican Hat III, Bright Angel, Joan, Sandra,* and *Bonnie Anne.* Five of the boats were marine plywood over oak framing. The *Camscott* was all aluminum, a prototype of a new fleet I hope to build; the *Bonnie Anne,* all fiberglass, was an alternate prototype built by a friend in Colorado and loaned for the trip.

Doug, after having shared the rowing of the *Norm,* now received a boat of his own, and four other oarsmen had joined the group: Don Ross from Bluff, Utah, Rich Chambers from Salem, Oregon, Dan Gernand from Palo Alto, California, and Chuck Reiff from Los Angeles.

Down through Labyrinth and Stillwater Canyons, each of the boats was a conversational island moving with the slow current. On the *Norm,* we talked river lore, politics, and told dialect jokes. From the *Camscott* came phrases like "delta foresets," "channel sandstone," and "geomorphology." Aboard the *Joan* they seemed to be talking recipes and cooking. One day the *Bright Angel* and the *Joan* drifted together for an hour or two and held a songfest, accompanied by kazoo, comb, and spoons. During an extended noon hour another day, Keri gave Rich an advanced course in needlepoint, which he works at aboard jets as a relaxation from the pace and pressure of his job. As they say today, everyone "did his thing."

Six days out of Green River we arrived at the confluence. It had rained the night before, leaving the half-sand, half-silt beach a soft, mucky surface in which our feet made ankle-deep tracks. But there was little concern with the mud underfoot, or even with the dark, foreboding clouds overhead. After six days of beautiful but slowly traveled canyon we were hungry for the fast water that now waited just ahead in Cataract Canyon. Despite the certainty of rain, a light demeanor — almost a party mood — could be felt all through camp.

Keri changed to wildly flowered, bell-bottomed hostess pajamas and, barefoot in the mud with a canned martini in one hand, began reading palms. She immediately became the kaleidoscopic core of a gathering of passengers and boatmen. "You like girls . . . but not extremely well . . . and you're tight with money . . . and stubborn . . . and artistic . . . ," she was telling one of the oarsmen as I joined the group, ". . . and you have a long Life Line."

Our spongy camp beach was a spacious quarter-circle sandbar where the right bank of the Green wheeled to become the right bank of the conjoined rivers. Behind the sand a hub of tangled tamarisk made a tighter arc against a steep slope of limestone slide-rock, and the sand tapered gently from the trees to the river all around the great curve. High above, just across the Colorado, druidlike sandstone shapes of the canyonlands seemed to be leaning intently over the canyon rim scrutinizing us, their heads and shoulders nearly enshrouded in leaden rainclouds. The evening was cool and the rain imminent, and nearly everyone hastened to make up a rain bed while dinner was being cooked over driftwood. As we ate, we watched the sky, thinking it might somehow brighten into a roseate canyonlands dusk, but it only became darker and heavier. The oarsmen who had "Green Glove Detail" for the day donned their rubber gloves and ran the cups, plates, and flatware quickly through boiling water, then secured the kitchen area against wind and rain. Everyone studied the surface of the beach until he thought he had found the spot least likely to collect water. Families laid out their bedrolls as deliberately as if they were drawing wagons into a circle against Indians. The bachelor oarsmen, characteristically given to seeking far-removed parcels of beach, clustered their bedrolls closer to the boats. But by morning none of the deliberations or preparations had mattered : everyone got wet.

The rain began at midnight, great heavy drops plopping on the plastic. Awakened by it, I pulled the wrapping closer around me and then finally, becoming accustomed to the staccato, drifted back to sleep. After an hour or so I awoke again, aware that something was wrong, and realized that a cold finger of wetness was invading my plastic cocoon. Groping at the stuff, I tried to find the opening the rain had found, but it continued pushing down through my sleeping bag like a mini-glacier.

The rain continued until daylight and by then hardly anyone was dry. Nor was anything in camp. Fred's fire was nearly impossible to restart, for we had forgotten to cover the driftwood pile, and only near the center of the thicker pieces, whittling away chips, did he find any dry ones for tinder. Finally the fire blossomed from a smolder into half-hearted, smoky flames, and immediately drew such a crowd that Fred could hardly work his way through with the coffeepot.

Still, the conviviality of the preceding afternoon and evening had somehow survived the night. The Jelkses, Womers, Philipses, and McKennas seemed in especially good spirits, and Malcolm keynoted the soggy but giddy breakfast gathering by dragging thirty pounds of sodden bedroll across the beach to the fire chanting, "I want a refund; I want a reee-fund," and grinning mischievously. We crowded around the fire for an hour or more, hot cups of coffee in hand, trading tales of our night in the rain, treating it almost without exception now as experience rather than ordeal. At the confluence, the distinction between experience and ordeal was easier to agree on than it would sometimes be in the three hundred miles of river still ahead.

THROUGH SATAN'S GUT

Our two days in Cataract Canyon could be recorded in a journal entry like this:

Sat. 19 July. Day 27. Mile 216.5R to Mile 202.5L (Confluence to Big Drop). Rain overnight. Beds rolled wet. Left camp 0900. Sky cloudy. Order: *Norm — Bonnie Anne — Camscott — Bright Angel — Sandra — Mexican Hat III — Joan.* Stops below third rapid (L), across from Turner inscription (lunch), at head of Mile-Long (R), and above first rapid of Big Drop (L). *Sandra* wrecked. Forced camp on tiny beach (L) ¼ mile above Satan's Gut.

149

Sun. 20 July. Day 28. Mile 202.5L to Mile 185R (Big Drop to upper L. Powell). *Bonnie Anne* capsized in Satan's Gut. Stops at 201.6R (f. aid), 198.5R (lunch), head of Gypsum (L) and Clearwater C. Don leads Mile 201.6–Mile 196.6. Dead water begins Mile 194. Row 3 : 00–6 : 00.

That is enough to preserve the bare bones of the adventure. But such a worthy adversary deserves more than a hundred and fifty words of summary recording that we "came, saw, conquered." Cataract Canyon did its best to give us hell.

Moist gray clouds still sagged low over the confluence of the rivers as we rolled the heavy, wet bedrolls and crammed them through the hatch openings into the boats. Clammy sand clung to the outside of the vinyl groundsheet enshrouding each sleeping bag, and the thought of having to crawl back into those same soggy cavities twelve hours later was undoubtedly in everyone's mind. Had the deluge come two or three nights earlier and a bright day followed, we might have taken a few hours to dry them and have made up the lost time. But there was no drying power in that day's sunless morning, nor certainty, yet, of time to spare. Only three days remained to challenge forty cataracts and row upper Lake Powell to Hite.

Through Labyrinth and Stillwater Canyons we had been a loose flotilla. There had been no danger of being swept into trouble and the boats had been allowed to spread randomly apart on the quiescent current. Each of them had now and then been in the lead position as oarsmen or passengers rowed, or had lagged behind with oars shipped as everyone swam alongside. Order and interval hadn't been important for a hundred miles. But all that was superseded at the threshold of Cataract Canyon.

Now the flotilla became an assault unit. I ordered a column of seven, with each oarsman responsible for maintaining a separation of four to five boatlengths from the boat just ahead of him.

The purpose of the column formation is to pass a workable course and its necessary maneuvers back from the leader to the last oarsman. A column of boats, with safe intervals between, extends far back upriver from the leading boat; it is often out of sight of oarsmen far back in the column, especially where the canyon is contorted or brinks of rapids can't be seen beyond until they are reached. This makes it impossible for the leader to lead all members of the column. He can only effectively show the course he has chosen to the oarsman just behind him, who by maintaining the right separation will be near enough to see, follow it if it works, and instantly improvise another — or head for shore — if it doesn't work. The column formation makes each oarsman in turn a follower, then a leader of the boat just behind him.

Sometimes precision is lost in the boat-to-boat translation. If a maneuver is made sooner, or later, or stronger, or weaker, it will have a different effect, and this may be passed back. Whether this variation is corrected or amplified in the translation depends on the proficiency of the next oarsman to receive it.

I had worked out a specific boat-boatman sequence for Cataract that I thought would best utilize the individual levels of experience at pulling oars and reading water, and familiarity with the canyon. The differing weights and loadings of the boats, and the size and muscularity of each boatman had already been taken into account at Green River in assigning boats.

First: The *Norm;* the proof-boat.

Second: Don Ross, with *Bonnie Anne.* Don had run Cataract many times and has his own vivid mental film-strip of it. If anything happened to the lead boat, Don would become the "guinea pig."

Third: Fred Eiseman with *Camscott.* Fred was still second-in-command of the expedition. If Don's had to become the experimental boat, Fred's precision with oars and skill at reading water would enable him to judge the success of Don's encounters and lead the remaining boats accordingly.

Fourth: Dan Gernard with *Bright Angel.* Dan had developed an almost immediate exactness in use of the oars. I wanted him to keep that precision by emulating Fred's, while being in position to study Fred's reading of the river.

Fifth: Rich Chambers with *Sandra.* Rich was good at both the use of oars and reading of water, and had been through Cataract several times in pontoons. He could either follow Dan, or look ahead to translate Fred's maneuvers if need be.

Sixth: Chuck Reiff with *Mexican Hat III.* Chuck is brawny and quick and was becoming good with oars. Following Rich, he would understand why a particular course was chosen, and find it.

Seventh: Doug Reiner with *Joan.* Doug hadn't known it, but his "running tail" was the reason I'd been priming him down through Whirlpool, Split Mountain, Desolation, and Gray Canyons. The last oarsman is subject to the most mistranslation of course and maneuvering, if there is any, by each succeeding oarsman of the column. He's the troubleshooter, charged with seeing that

The figures that inhabit Utah's Land of Standing Rocks seemed to peer over the rim at us as we approached Cataract Canyon.

all other boats are finding safe passage down the river ahead of him. Fred and Doug were brackets of proficiency between which I had enclosed all except my *Norm* and its potential replacement.

As we pushed off from the big confluence sandbar, the scene was almost a caricature of gloom. The fawn-colored walls of the gorge were trying to be gray like the morning. A layer of damp clouds was draped from the top of one cliff to the other, like a dingy cloth, and a small rip just over our heads taunted us with the probability that there was blue sky for everyone everywhere else, but none for us. More than an inch of water was already in the boats from the rain, and muddy sand from our feet. The life jackets were cold-wet. Very theatrical; what more could one ask if one were staging mood for an entry into the canyon known as the "Graveyard of the Colorado"? Obligingly, I even had the "stage fright" that wells up when there are rapids ahead, and needed to get down onto the tongue of Number One so everything would be real again.

Like the Green in Lodore, the Colorado in Cataract runs calmly and rather straight for its first four miles. Then it veers impetuously as if to put all boatmen on notice that a change in disposition is forthcoming. But the rowboat oarsman, unfoiled by motor noise, is already on notice. The voice of raging river has already reached him, far upstream from the canyon's hard turn to the left.

Fifty yards above the head of the rapid, I judged the limit of prudence had been reached, and reluctantly put on the clammy, rain-soaked life jacket. Looking back quickly, I saw that the signal had ben anticipated; some had already finished with theirs and others were snapping fasteners and tugging at D-ring adjusters. By then, standing at the oars, I could see the tailwaves of Number One beyond its flat, glassy brink. It was short, apparently a moderately sharp drop. Waiting for a closer look, I wet my hands and rubbed the oar handles.

At the brink it looked clean down the middle, with a decent tongue running into a procession of tailwaves and rocks along the sides of the tail, but none blocking the channel. With the point picked at which to start down the tongue, and the stern downstream to meet the turbulence, there was little more to do but ride it out. The *Norm* hung almost imperceptibly on the threshold, then slid down the smoothness into the converging lateral waves, and through the bouncy tail. When the waves had lessened enough to let me look back safely, the *Bonnie Anne* was back in the biggest of them and the *Camscott* was poised, about to enter. The others were still out of sight beyond the top of the rapid.

There was perhaps a half minute to watch the *Camscott* plunge through, and the *Bright Angel* take its place at the brink. Then the diminishing voice of Number One had been overpowered by the rising voice of Number Two, compelling me to look. It was choked with boat-sized rocks and it took a few seconds to see a clear entry on the left that led down opposite the head of a clear running channel near the center. To make a continuous course of the two I entered the left channel pulling hard to hold back, increased my quarter as I moved down and to the right behind the center rocks, then let the current have us when we had headed the center channel.

"How's Don doing?" I barked back toward Joan, behind me in the bow seat.

"He's right with you."

"See anyone else?"

"Not yet . . ." Then: "Here comes Fred."

I hoped everyone had kept close interval so the change of chutes in mid-rapid could be seen and passed back from boat to boat. But we were moving down on Number Three.

Three was like One, fairly clean and uncomplicated. We shot it down the middle and caught an eddy against a left-bank beach just above the head of Number Four. Three rapids in six-tenths of one mile; time to stop and study the reactions of the individual oarsmen. The last boat, Doug's, was just coming through Number Three by the time I'd finished tying the *Norm*. Don's had touched the beach beside me, and those between were in different stages of pulling from the fast water into the eddy.

Pre-rapid tension has several effects, one of which calls for the implementation of the "women upstream / men downstream" rule more frequently than in tranquil sections of canyon. The other is to make the oarsman less communicative to his companions. Most of his communication is with himself and his river. When the tension is high, the garrulous become close-mouthed and the quiet become nearly antisocial. But everyone was talkative at the foot of Number Three. We had quite literally "gotten our feet wet" — and other basal parts of our anatomies — and we stood around the bows of the beached boats chattering like Monday morning quarterbacks rehashing the weekend games.

The rest of the morning was as delicious as the beginning of Lodore had been. All the rapids were low on the rating scale — twos, threes, and fours — only a fraction of a mile apart with fast river between, and we ran them "wide open" by standing, studying a few seconds, picking the point to go swooping down the tongue, and plunging through the tailwaves. We made no more stops for almost an hour. By late morning the gray clouds had stopped their threats and given the sun to us, as well as the rest of the world, and the map indicated we'd run through Number Ten.

On other sections of the river, some rapids are named and located by their milepoint, but Cataract's rapids are so close together that using milepoints to locate them would require decimals rather than the nearest whole number. For example, the first three are at Miles 212.3, 212.0, 211.7, with only three-tenths of a mile between their heads, and less than that between the tail of one and the head of the next. Boatmen try to designate them, instead, by numbers: One, Two, Three, but the river evades this attempted mastery by varying the number of rapids with different volumes of flow. Higher water pushes through to run contiguous rapids together into longer, fewer ones; lower water again lets them separate, with only fast current or perhaps even casual water inbetween. A numbering is good only for a certain latitude of flow, and no two boatmen's numberings seem to agree. Cataract's *pièce de résistance,* the wild water known as Satan's Gut, is variously labeled as low as Number Twenty and as high as Twenty-six. So it is useless — and foolhardy — to tally rapids as a means of keeping track of one's position in unfamiliar canyon.

The count, then, didn't matter. It was more relevant that I keep the strip map handy to match river bends with directional trends of the canyon seen ahead, and that Don was right behind to take the other boats where he thought they should go if my course got me into trouble. We had already guessed that the medium-low stage of river we were on might allow running as far as the head of Mile-Long with just careful listening and quick reading from the brinks, and it did. That made the first five miles wonderful sport tempered only with the usual readiness to be surprised. I'd have delighted in running threes and fours like those all day, with some precocious fives thrown in just to keep me keyed up — as long as we hadn't yet come to the prominent left turn at 205. That would plunge us into the head of Mile-Long and just below, the Big Drop.

When I could look back and see the other boats, our column formation was like a caterpillar snaking its way through the canyon. A movement around a midstream rock, or laterally to avoid an eddy, would flow back through the creature, the reaction of one boat being imitated by the next, and the next, and the next, as if an invisible nerve cord connected and triggered them in turn. The skill and coordination of the oarsmen made it seem so. Yet each segment of the caterpillar was its own body and brain, with ears and eyes and stiff wooden flagella to stroke the water. Cut it anywhere — even into seven — and each part would continue breathing, listening, looking, thinking, stroking. It was a beautiful creature, it always is, because it is smarter, more capable, more likely to survive, than any of its segments alone. Watching its skillful articulations always gives me little shivers of pride.

Just before noon, Joan sighted a name we had been looking for on the right-hand wall. It was Turner, either Al or A.G., and had apparently been done with black paint and a brush or driftwood stick. It probably dates back to 1907; the Kolb brothers noticed it in the fall of 1911 and later met a man at Hite Crossing whose friend Turner had successfully negotiated Cataract several times. There are many names inscribed on the walls of the Graveyard of the Colorado. Safe passage through

Cataract has historically been manifested by the leaving of inscriptions after the worst rapids are passed. Other of the canyon's challengers, perhaps uncertain they would survive to congratulate themselves below, have entered their names nearer the beginning than the end. Our purpose in locating the Turner inscription was to photograph it for "Dock" Marston, a Colorado River historian.

The *Norm* was in the tailwaves of a small riffle when Joan pointed out the black lettering. I finished running it, then caught the edge of the eddy between the riffle and the right bank and rode the eddy current back upstream abreast of the waves to put the name squarely in front of the camera. That fulfilled our photo assignment, and two or three more boats had by then entered the eddy to wait. We cut into the edge of the riffle high up, ran it again, crossed, and landed for lunch.

The shoreline of Cataract, especially the left one, had been to that point a series of beautiful sandbars, spacious and untracked. Temptations for an early camp were strong at each; I could see where we'd tie, where we'd get firewood, where I'd scoop out my bedroll pit. But the sun was still high and hot, and too much of Cataract lay ahead.

Not far downstream, the canyon started its big warning turn. Within a few minutes of leaving our noon stop we could notice the right-hand wall swinging strongly left. The river, though, contented itself with an urgent but smooth flow, the declivity there being five feet per mile. At Mile-Long Rapid it would start dropping more than thirty feet per mile.

As it turned out, the view from the right bank at Mile 205 did very little except pump a little more adrenalin into our bloodstreams. The Colorado scudded around a jumble of boulders that narrowed it severely, growled at them, and rushed on to roar in earnest below. There it piled hard against the right cliff, then cut back out of sight to the left. A rapid — even a mile-long chain of five or six rapids — is possible to reconnoiter if there is a shoreline as there had been at Disaster Falls. But a rapid that pitches down and down and down over rocky ramparts and simultaneously writhes in its harsh bed to demolish its shoreline is another thing. Mile-Long, for the hundred yards we could see, had left only scraps of shore, first on one side, then on the other, with no way to see it all except to run it. The long rapid was to us a Great Unknown, as the entire river had been to Major Powell. I sensed a danger, at least to myself, in thinking too long about it, and headed back upstream over the rocks to the *Norm*. Don thought our stage of water *might* leave a few yards of slower river — possible landings — between the cataracts at two points.

Being unable to study Mile-Long and plan the run, I had no picture of it in my mind against which to project a little *Norm* like a model warship on a tabletop ocean. No way to see its position and where it had to go next, nor, afterward, any feeling of having run a visualized section of canyon. Only of running a wild medley of waves and rocks. I don't think there *were* any ordinarily structured rapids with smooth swooping tongue and bordering lateral waves converging into a tail of diminishing wavestacks. Only immense rocks fallen in from both sides and the river confounded and angered and dashing back and forth at them and raging over and around them in obsessed fury and the *Norm* finding a way through it all.

The canyon had walls, I know, but I didn't see them,

and I suppose there *was* some quiet water in there somewhere for a panic landing, but I don't remember needing it, and not needing it, didn't want it. I wanted only
to slip past the next rock so I could dip hard and deep
behind it and pull, or for a millisecond to pass so I could
knife the oar blade into a wave's fat crest rather than
its meager trough. Ragged waves knocked down by the
stern crashed onto the deck to be split by the pointed
splashboard and knocked sideways, back at the river.
Others heaved sloppy crests over the oarlocks into the
cockpit. We picked our way around rocks treacherously
hidden under too little water and thrashing holes that
could have seized and held a boat forever. At one point
I think I somehow crossed nearly the entire width of the
river in mid-rapid to get to a clean chute on the other
side. Over all was the awesome thunder of a mile of
wrathful river. Even within the *Norm* we couldn't talk;
we could only throw our voices against it. "I can see
one boat," Joan shouted, or ". . . two," as I was running one rapid, looking ahead toward another, speed-
reading it, deciding it said Go, going.

Somewhere in Mile-Long I found time to glance back
at Don for two or three strokes, to know how busy he
was, and how deliberate. His boat was part of him; he
was unaware that the oars were wooden and separate
from his hands. His arms and shoulders were taking
care of things in one boatlength while his mind resolved
the one ahead and his eyes searched out the one just
beyond that. Surely he was feeling the same precious
fear I was. And what of the other five I couldn't see?
Were they feeling it? Was it working for them or against
them?

Suddenly, the other world existed again, the world of
sky and cliffs and awareness of the colors of things, and

161

flat water ahead of the stern. A quarter of a mile or more, with nothing happening. A chance to recover. Time to unsnap the bucket and bail, get the boat lightened; there was ankle-deep water in the cockpit. Time to pivot a quarter-turn and start counting boats behind: three . . . four . . . Joan was bailing the bow end with a tin can . . . five . . . six — all right side up and no hand signals of any kind from them. All okay — seven for seven! In Mile-Long we made no mistakes. Or if we did, some part of our seven-bodied creature intercepted them and set them right.

Bailing and respacing as we drifted, we came down toward a boulder island in mid-river. I could see no self-evident course, but with perhaps ten boatlengths to go knew that little could yet be decided. Either the left channel or the right channel around the island would in time present itself as the better of the two. But then it didn't. Both divisions of the split river seemed to be equally wide at their heads, to be receiving equal amounts of water, and to be roaring identical roars from their angry sections down out of sight. With five or six boat-lengths to go, I still couldn't decide, and looked back at Don. He understood the question in my glance, but spread his hands palms up and shrugged his shoulders. Four boatlengths. If there'd been only the *Norm* I could have held above the head of the island on the fine line — like the part in one's hair — between the right-moving water and that starting left. Or I could even have beached there at the precise division of the water and walked down the island for a look. But there was no place to hold or land six other boats, which is perhaps the Achilles' Heel of a creature having seven bodies to take care of. With a few feet to go before touching boulders, I quartered and pulled to let the right-thrusting water have us and we veered away and down. The swell of the island narrowed the chute, deepening it somewhat, which triggered a little surge of optimism, but then I saw that the entire bottom of the chute from the island to the right riverbank was crossed by a barricade of long, rough waves.

We were centered in our chute, stern down, and prepared again to move quickly either way, but nothing looked good. It just didn't read right: too regular. My mind sorted through all its pictures of rocks and water, looking in its own dark corners for something it used to know. Water did that sometimes. When?

Suddenly, there it was, an old mind's-eye picture of a place in Glen Canyon that used to look that way in September and October. Bedrock! That's what it was! Ledges — upthrust ends of rock layers like the tops of leaning books, tearing at the river's underside.

Perhaps I pulled left, close to the island's bouldered edge, because I'd found a few extra inches of water over a similar ledge against a similar island in the days of the Glen. Or perhaps there really was nowhere else to go. There seemed to be a smoother slot through the corduroy of harsh waves there, where the *Norm*'s left oar barely cleared the boulders, but sliding into it we rose on a crest, then grazed something — rock or bedrock — in the trough below. Was it *the* place to run? I still don't know, and I'd like to. The river usually gives you a more definite yes or no.

The channel left of the island couldn't have been run at all. It might nearly have been walked. Looking back into it from below, we could see that it consisted of a long terrace of smooth water coming partway, then abruptly brinking and dropping thinly over bouldered

163

terraces. From our perspective, that half of the river looked like a misplaced creek running down stony steps, steps that deserved marble lions, or gargoyles, set at either side.

Don, then Fred, then each of the others, seeing that the little chute was at least a workable passage, lined themselves up and also scuffed through without damage. Ahead on our right a tributary canyon entered. "Calf Canyon?" I called to Don. The deliberate extravagance of his yes nods not only answered, but conveyed the rest of the message the map was already giving: the start of the Big Drop. Of the Graveyard of the Colorado, the Big Drop is certainly the cynosure. The cataracts are its catacombs, the dreadful holes its crypts, the mid-river boulders its headstones. It was time to begin picking our way.

The boats were landed slightly upstream from Calf Canyon, on the opposite side. A long sandbar reached down toward the head of the rapid, and slanted up from the river to jumbled rocks overlooking it. The river had at some time been about thirty feet deeper and had left a mantle of driftwood and flotsam the length of the sand just below the rocks. It was rather steep, but a nice beach. Little did we know how much we'd be wishing for it later in the day.

By going down the beach a short distance and then up onto the rocks overlooking it, we could see the length of the rapid. It had two parts. The upper was a mild tonguing caused by projecting rocks on both sides opposite each other. The river pushed between these with very little protest, but just below was intercepted by a promontory of big rocks tumbled in from the left bank that threw it back to center. Beyond that, its tail was an angry row of ragged stacks gradually absorbed by slower water below, and as the river went out of sight to the left, it was calm again. The tailwaves looked rather wet to run, which was a consideration at that point. If we took too much water on board, the boats might be too heavy to pull in above the next rapid. The tailwaves would have to be avoided, if possible. The danger was that in avoiding them, we might be bashed against the rocky promontory. I suggested that the oarsmen and I walk farther down to see how much room actually existed between the tailwaves and the outermost rocks.

From our previous overlook we had been able to see two huge rocks as the outliers of the point. From our new one we could see several smaller ones huddled around them like helpers. They were black with time and water, like battered rocks of a seacoast. Out beyond them, nearly concealed by water skidding over it, was another. From a boat, it would be undetectable until the last few feet. We call them "sleepers," those nearly hidden boat breakers. That particular one further narrowed the passageway between rocks and tailwaves. "We'll have to play the run off that sleeper," I cautioned. "Better, even, to run right down the middle of the tail than get tangled up in those rocks."

We decided to enter the upper part of the rapid well to the left so as to be out of the line of the tailwaves. As we approached the promontory of rocks, we'd move right just enough to pass outside the sleeper; that would clear it and still give us maximum distance from the chain of waves. Before starting back to the boats we visually measured the distance from the visible rocks to the sleeper, and turning to restudy from several upstream points, we gave ourselves a good idea of what the sleeper's covering wave would look like as we approached.

I had made my run, having to quarter to the tailwaves and pull to stay out of them, and had landed on a splinter of beach with the other boats coming around the promontory in order, when someone from one of them began bellowing, "The *Sandra*'s holed! The *Sandra*'s holed!" Whirling from my tie-up I saw that she was just coming in right side up, with everyone aboard, but her stern was deep in the water. Her portage bar was under, which meant she was drawing about six inches too much, and Rich was having trouble making her respond to the oars. Each powerful pull gave only slight progress toward shore. Fortunately the rapid just below had a damming effect on the river so that its slowness gave him time; there wasn't much chance of being carried over it, as we had thought above Calf Canyon there might be. But it did appear that her stern might go under before Rich got her in. He was not only having to cross half the river to land, but was having to row quartered somewhat against the current so that he wouldn't be carried by before he could land with the other boats. After what seemed a very long time, the *Sandra*'s bow touched the sand, her stern deck nearly awash. Immediately a squad of boatmen and passengers moved into place around her. "Let's get her up as high as we can," I told them, and counted "one-two-three-HEAVE!" several times, but could only gain a foot or two up the gently sloping shore. Even that wasn't enough to put her usual waterline high and dry.

Rich opened the stern hatch, and he and Chuck began dragging gear and supplies from the elbow-deep water inside. A human conveyer was formed to pass each piece from one person to the next, back onto the rocks behind the little beach. The sleeping bags, completely saturated, were almost too heavy to lift. When the largest items were out of the stern compartment, a bailing bucket could be dipped through the hatch opening. But quite a few buckets were dipped and thrown back into the river before we realized the water level inside the compartment wasn't being lowered.

With the sodden beds and heavy rubber duffel bags removed, the *Sandra* had been lightened just enough that we could "one-two-three-HEAVE!" her farther up the bank. Most of her hull was then out of the water and I could see a large hole at her chine, where the side and bottom join. Before I could get down to examine it, someone on the other side said, "here's the hole," and I went around and looked and found one there, too. Somehow both sides had been hit. Each hole was the size of two fists, doubled, and each was profusely pouring water from the boat back out onto the sand.

When the river water had run out, we dragged her farther up so that the stern was almost clear of the river and the stem was against the rocks that truncated the narrow little beach. Then we turned her over so the damage could be studied. The hole through the left chine was less critical because it had broken into the bottom of the sealed boatman's seat, which doubles as a flotation chamber. The other hole was much more serious because it had broken into the stern storage compartment which is a single cavity about five feet in length and in width, and averaging fifteen inches deep. It is the main cargo hold, but more important, is the boat's largest flotation compartment. The buoyant heaviness that characterizes Cataract Boats is due to their seven watertight compartments, of which *Sandra* had only five remaining undamaged and those comprised less than fifty percent of her compartmenting in volume. Unless the hole into the stern could be patched, we'd have to leave her lying there broken on the sand to become more bones for the Graveyard the next time the river rose. Our patch would have to slow the leak enough so that water could be bailed faster than it came in. With a thousand pounds of water on board she would draw well over a foot and be impossible to maneuver; she would almost be part of the river.

One couldn't help feeling trapped there, between two rapids of the Big Drop on a scrap of beach twelve or fourteen feet wide and sixty feet long with river in front and rocks crowding behind and twenty-three people to be accommodated. Somewhere very near — we didn't look for it — there was supposed to be an inscription that said:

CAMP NO. 7 HELL TO PAY

SUNK AND DOWN

and my own disquietude was as well told as it could be, by those cryptic words of someone else in trouble there once. It was our seventh day, too.

The rock that had broken into the stern compartment had hit squarely on the chine, punching away curved pieces of both sides and bottom to make a hole almost oval-shaped. It had also torn away a section of the mahogany chine piece, which fastened the side and bottom of the boat together. There was a lot of wood to replace, as well as waterproofing to do, if she was to float again. But it looked as though she might. I went to the *Norm* for my tools and precut squares of patching plywood.

"Can you fix her?" someone asked.

"I think so."

"Then is this camp?" asked someone else. I'd been fear-

The Sandra *below Mile 203. More bones for the Graveyard of the Colorado?*

ing that question and the reaction to my answer. There was hardly enough room for everyone to stand on the stingy sliver of sand, let alone our rolling out bedrolls and building a cooking fire. In the back of my mind, perhaps I was expecting our passengers to act, as two or three had in years past, as if these anguishing boat wrecks were be to staged as part of the adventure, but of course, without any inconvenience to anyone. I should have given this group more credit, and would have if I'd been thinking about their cheerful adaptation to the previous night's rainstorm. Everyone pitched into the making of camp and the repair of *Sandra*. Joan and Priscilla remembered having seen fragments of a broken foam ice chest among the flotsam at the head of Calf Canyon Rapid, and offered to work their way back up over the rough rockslides to get them for possible "wadding" for the two large holes. Rukie Jelks took his boys and went looking through the rocks for a piece of plank or timber from which a new chine-piece section could be sawed. As the work went on, the men were always there when *Sandra* needed turning from one side to the other to enable my working both inside the hull and out. Maggie and Keri helped scatter wet gear from the stern compartment over the rocks to dry. Everyone spread the still rain-wet beds to dry in the sun. Having such helpful companions in misfortune made the problem seem almost routine and my task a much less lonely one.

As had the other boats, *Sandra* had known her share of bumps and bruises. Her chines, and her sides and transom had been broken before, opened up to the river so that moisture had unscarfed and delaminated some of her marine plywood, and the fiberglass skin was probably what preserved her rigidity and watertightness. My father, a craftsman with tools and materials, had always been able to tighten her up. "This season's her last," I'd promise him every spring, but I always needed her again, or was unwilling to block her up on sawhorses and abandon her. There's a parallel in a story I know: Once, out on the Utah desert, there was a little religious sect. The day came when one of its members died. Her associates, not believing in death, dressed her in finery and seated her on a chair in the communal building. Every day, so they say, her hair was combed and she was talked to and given nourishment somehow. For a long time. Until a rancher living downwind complained to the authorities. The *Sandra* was like that poor, dead woman who had to keep fulfilling a role, long after it should have been expected of her.

She had been beached on her right side, so I decided to work first on the left. This entailed chiseling away splinters of broken wood to make a clean hole, and rasping a roughness on the surface of the surrounding fiberglass skin so that a fiberglass patch, to be applied last of all, would adhere. The first patch was placed over the portion of the hole that was in the boat's bottom. Nails were then driven through the patch and a thick layer of tar beneath it, into the boat bottom. Next, a second patch piece was put into position on the *Sandra*'s side and slid so that its lower edge was tightly against the edge of the one repairing the bottom. That gave her a plywood "skin graft" of sorts, over the corner of the hull. Then to make sure we had the maximum watertightness, Chuck drove several screws and twisted the screwdriver until the black compound began squeezing out at the edges of the patch. The last step was to cut a piece of fiberglass cloth to size, stir some resin-catalyst mixture, and lay another skin over the entire wound area. That done, we waited a few minutes for the fiberglass to "go off."

The repair of the other side was more complicated because the hole was larger; more material was missing. Joan had found the foam, back at the head of the rapid, and through trial and error Rukie and I eventually trimmed a piece into shape and pressed it in to replace the missing portion of the side. For the bottom, I carefully cut a thicker rectangle of plywood to length that would just let it go down between two oak crossframes if I pounded it; then after coating the underside of it thickly with tar, drove it down. Its new stiffness supported the area around the hole. Against the broken side of the boat, inside, I nailed another tarred patch, clinching the nails outside the hull so they'd hold, and then fastened a sawed piece of driftwood plank into the corner against the patches, as a new chine section. A fiberglass covering completed the job. By that time it was nearly dark.

I decided not to put her afloat again until morning, much as we needed her place on the beach, because the overnight hours would harden the fiberglass more gradually if it remained warm and dry. As a compromise, we left her resting on her side, braced with oars against wind gusts or settling sand. Beside her, underneath each of the two brace oars, room for a bedroll remained and by making a mosaic of other bedrolls we found room for nearly everyone. Four of us found a steep watercourse just beyond the end of the bar in which large rocks had trapped little terraces of gravel behind them. By digging out the coarsest gravel hunks and filling their cavities with finer material we made relatively smooth surfaces on the most level of these. The dinner fire was made near the downstream end of the beach in a shallow but steep-banked gully that wouldn't accommodate a sleeping bag, and when we had eaten, we doused the embers and buried them in sand. After that I lay awake for a while in the sweet night air, my eyes on the stars and my mind on the river, on the twenty-seven days behind and the seventeen days ahead. Then, almost immediately it seemed, morning had come.

While the others were having breakfast, I made my way over the coarse talus blocks down to the head of Satan's Gut. The rapid just above it — just below our crowded little camp — looked runnable, given careful entry and some healthy trepidation about hitting a chute between loud, deep holes set partway across the foot. After that, quiet water pooled along for fifty yards to the brink of Satan's Gut. Don had said several times that "you *have* to run it down a slot on the left side," and now that its turn was coming soon I wanted to be thinking about the way it was structured.

At first glance I thought it was the worst piece of white water I'd ever seen, but then when I decided it might be runnable I had to give honors back to Crystal Rapid in Grand Canyon. Satan's Gut, when viewed from partway up the talus, looked like a narrower, more severe version of the dam we had run above Green River. The curve was not of smooth concrete, but of cruel, mighty rocks. The river above was nearly brought to a standstill by their close-set, steeply coursed arrangement from bank to bank. Then, when it finally pitched over the edge, the waiting rocks instantly tore it to shreds. Some remaining declivity below, mild when seen as part of the same structure, caught it still fizzing from its stupendous aeration, started it moving downstream again, and narrowed it into another rapid, a mild one, not far beyond.

I speculated that a wide rockslide must have once tumbled from the opposite cliff to actually dam the river and that only the stronger current of midstream had been able to push and tumble some of the great blocks away in the direction of least resistance, downstream.

Just out from the left bank, a pebble's flip from where I stood, was the feature for which the rapid is named. A strand of water emerges taut and glistening from a chink somewhere up near the brim of the harsh, bouldered face and stretches down to disappear into the featureless turbulence below. It is narrow and it shines and its slippery-smooth appearance sets it visually apart from everything else in the rapid. Looking down on it is almost like looking into an immense surgical incision. The name of the rapid is apt; its most prominent feature looks visceral. Down that glossy filament of water that was no wider than our boats we'd have to run to make it. Entered right it would amount to perhaps three boat-lengths run between bottomless, thrashing abysses, then safe water below. Entered wrong . . .

With breakfast out of the way we unchocked and unbraced the *Sandra* to test the effectiveness of the repairs. By carrying her coffinlike for a few feet, then setting her down in the water, we avoided any possible loosening of the patches from contact with the beach. There was no way to see whether the seat compartment still leaked, but inside the stern patch nothing appeared for a minute or two, then a few beads of water began pushing through and collecting. She would need periodic bailing but she was serviceable; she could go with us. Her crew began loading her, and soon all the oarsmen were standing at the bows of their boats, the signal they were ready to shove off.

To reach the head of Satan's Gut we first had to run

the rapid just below camp, which was more in the nature of two obstacles joined by fast water. We had studied them during spare moments of the *Sandra*'s repair, and again after camp was broken. Two large tan boulders protruded six or seven feet out of the water, one near the right bank and one near the left. This effected three chutes with the option of running left, center, or right. Each of the chutes, however, led into a cross-trough of holes and the only clear channel was directly below the rock farthest right. To reach it we had to enter the center chute and pass almost within touching distance of the rock, then quarter quickly into the quiet water in its lee and drift straight down from it. Below, fast water thrust into deep holes — the second of the obstacles — that nearly crossed the river, but there were navigable straits between them. All boats made beautifully precise runs through both obstacles. The nearly still water ponded behind Satan's Gut provided ample room to make our pre-rapid landing.

If we were to use that fascinating, glistening "gut" as a way down through all those rocks, we needed some way of finding it from upstream even though it couldn't be seen. Something that marked it as a safe zone to fall down onto when we put our seven boatloads of people over an edge that, until much too late, hid everything that was below it.

There are two ways of pinpointing an entry into a rapid that drops away so steeply. One is to use a signal man at the head. He must be very skilled at reading water. By standing high enough above, he can see the rapid, the entry point, and a boat approaching, and if an oarsman is going to miss his entry can give him hand signals for Right, Left, or Hold. Boats can be divided and run through in groups of two or three, with at least one other good signal man needed to guide the first when he brings the second contingent through.

The other method is to select a locating feature that can be seen and trusted at the head of the rapid, even though one can see nothing beyond. For Satan's Gut I had chosen a small "curler," a wave about three feet long, six inches high, with a half twist of white foam tending to fall upstream from it most of the time. It lay broadside to our intended course, its left end terminating just above the drop onto the "gut." The next step was to determine how to use the curler. We talked over the idea of playing our run off the wave, and made sure everyone's eye found it. Then, as the rest of us watched, Dan threw several large pieces of driftwood out at differing distances from shore, where they lazed along until they reached the edge. After they plunged over the brink most of them were never seen again. The only ones saved from perdition were those that went straight down Satan's Gut. They were the hunks that had been carried over the last two feet of the curler, at its left end. And so that seemed to be our safe zone: one-third of a pretty, caramel-colored wave with a white pompadour.

In the instant I could look down on the rapid before dropping in, I realized the water had carried my four-teen-hundred-pound boat a little differently than it had the arm-sized chunks of wood. With inches as important there as whole boatwidths would normally be, I was in the wrong place — we should be taking the middle of the wave instead of the left one-third! These realizations were projected against my stare down — down, down, down — into a monstrously thrashing void just three feet off my left stern quarter. There was, I know, one stroke, one deep, desperate holding stroke from my feet against the bulkhead up through pushing legs and taut stomach muscles and shoulders and elbows and then a prolongation of it by hauling the oars back just as far as I could lean. I may have gotten part of a second one like it. They would have been upstream strokes, weakening the right oarpull toward the end, to get both holding and right-quartering action. Satan's Gut was on my right, a half boatwidth away!

It will always be vivid, I think, the memory of that foam-filled abyss, and how we hung above it, and the stroke slowed us a little, and then a wondrous vagary of last-chance current skirting the brim to go down the Gut took us with it. We were in and then out of danger in perhaps five seconds, so that fear and relief must have arrived almost together if fear was there, for I felt little of any emotion and had only the awareness that we were making it.

Don was by then nearing the brink and having to make a snap decision. When he had seen me pitch out of sight still trying desperately for a move to the right, he reasoned the Gut was farther right than I had calculated and that I had missed it. He did what he should have done, and all he could do — he moved quickly right, perhaps a boatwidth more than he had been. But when, a few seconds later, he got his first look down into the rapid, he was about to drop into a steep brawling hole. Only then could he know that I *had* achieved the Gut — and that it was now on his left much too far away to hope for in two strokes.

Joan was watching from the bow seat of the *Norm* to report the success of the other boats, and saw *Bonnie Anne* hang on the brink, drop out of sight momentarily, then wash out of the hole upside down. Her "Oh — no!" shot iced lightning into my own guts and spun me on my

seat. A wild hell of foam and rocks and in it the unpainted bottom of the capsized boat, and somewhere,
Don, Win, and Yosh. It seemed forever until heads and
life-jacketed shoulders began bobbing in the spumy
waves around the *Bonnie Anne,* the waves never bringing all three into sight at once so that I could tell myself
the worst was over.

"Guide me," I yelled at Joan and began pulling long,
holding strokes with the bow directly against the current so that everything and everybody would gradually
catch the *Norm.* "Can you see them all yet?" I called
over my shoulder after several strokes.

"I see Don and Yosh. They're getting up on the
bottom."

"Keep looking." I needn't have told her that, I realized
as I said it. But where was Win? Perhaps if she looked
hard enough for him, she could somehow make him be
there.

By continuing to slow our downstream speed with
the oars and having Joan tell me whether to move left
or right to put the *Norm* directly below her, I got the
Bonnie Anne close enough that we could take her bow
line from Don. He and Yosh were on their boat's bottom.

"Where's Win?" I implored.

"I think he's hanging on to the *Camscott*," answered
Don. "He ended up closer to them than to us."

"Thank God. I'm going to try to tow you into an
eddy." Then to John Womer on the stern, "Get a good
grip on that bow line, and brace yourself."

As I resumed pulling against the current and quartered to the right bank, the *Bonnie Anne* became a
pendulum on her long nylon line, swinging past to
precede the *Norm* downstream. Now the combined load
was about twenty-five hundred pounds and I couldn't

muster enough leverage from my oars to move it more than a few inches toward shore at a stroke. We were being taken inexorably down to the next rapid, which showed a large rock in the center at the head, but looked to be clean below. I decided the only thing to do was to give up hope for the moment of getting the *Bonnie Anne* to shore, and instead, to let her down into the rapid, clear of the rock. This we did, it fortunately being mild and clean so that it presented no danger to the men riding on the bottom. When smooth water was achieved again, we took up a few feet of slack from their line, which we had continued to hold, and resumed our efforts. Within a few hundred feet we found an eddy we could gain with our awkward lash-up, and worked it to the right shore. The other boats were soon there.

Don was bleeding badly from a cut below his left eye. Win came in the *Camscott*, very white and having swallowed an appalling quantity of silty water. Yosh had scraped shins and was sure his Cinemascope camera had gone to the bottom. We had been on the river less than ten minutes.

Rather against his wishes, Don was made to lie down on several life jackets and a shade was improvised over him of oars and groundsheets. Malcolm, from his knowledge of anatomy as a paleontological tool, called me aside. "I couldn't say for sure," he cautioned, "but Don's zygomatic arch may be fractured. His cheekbone. From the way that wound is depressed, I'd say it is."

Keri had already foreseen the possibility of shock and had opened a bedroll to cover Don with the sleeping bag. She, Maggie, Priscilla, Mary, and Joan had taken over the cleaning of his wound and the application of "butterfly" bandages to pull and hold it as well closed as possible, and later handled the spreading of his gear on the rocks to dry. He being in excellent hands and joking that he loved being surrounded by beautiful women but didn't like what he'd had to do for it, the rest of us went to get his boat squared away for him.

The *Bonnie Anne* was snubbed against some boulders in about three feet of quiet water. Borrowing a bailing bucket, I put it upside down on the water beside her and pushed it under to trap the air it held. About a foot down, by tilting the bucket inward, I could release the air into the inverted cockpit, in which some air was already trapped. Through repetition of this procedure several dozen times, buckets of air were bailed under the boat and caught in the cockpit, to raise the *Bonnie Anne* several more inches by displacing more water. This made the next step easier. We recruited every willing passenger and every boatman except Don and had them reach under water and take hold of the safety rope that runs along the side. On the count of "one-two-three-HEAVE!" we hoisted the side high into the air, then pushed it from us, so that she rolled on over into a righted position.

Even without knowing she had come down over rocks after capsizing, one would have known that some harsh and merciless force had raked her topside. The vee-shaped aluminum splashboard just ahead of the cockpit, which was her deepest point inverted, had taken the brunt of the damage. It was wiped askew and flattened nearly to the deck and had deep scratches from the coarse rocks. The bow splashboard, of $9/16$-inch marine plywood, had been torn away altogether, taking with it the spare oar attached by a quick-release device. Neither the oar nor the splashboard were found; the weight of the big iron tholepin ring had overcome their buoyancy and taken them to the bottom. The active oars had been

saved by their strong nylon safety cords. The load inside
had fallen down on bow and stern hatch covers and dis-
placed them from their seals to admit all the water the
trapped air inside would allow.

Once again, as we had a few hours before, we formed
a human conveyer line and passed each soggy article
along it to be opened and spread in the sun. One of the
first pieces encountered, much to our relief, was Yosh's
camera. It had been in a compartment after all, rather
than under his cockpit seat as he feared, and the buoyancy of its watertight Halliburton case had held it
tightly there.

Don went into a mild shock but, I think, held off the
brunt of it through sheer guts and determination. Un-
compromisingly conditioned by having led trips of his
own since his middle teens, he was simply unable to give
himself the luxury just then. You don't let up on yourself until everyone's off the river, and however much he
may have wanted to, he didn't. He allowed himself the
shade of our lean-to but not the indulgence of a catnap,
or of even lying fully prone: "No, I'm okay. I really feel
pretty good." We allowed him a cigarette or two, which
may or may not have been prudent, but was perhaps
better for his morale than the denial of one. Within an
hour he told me in a casual way, but with much inferred,
that he should "get back on the river pretty soon." I
knew full well what he was really saying. I've said it
to myself: the whole river was becoming a Satan's Gut
in his mind and he sensed that and had to get back out
in real tongues and real tailwaves and stop replaying
the accident over and over.

"I'd like you to lead for a while, if you feel up to it,"
I said to him a half hour later as we finished reloading
the dried gear. I'd rehearsed the words, the way of

making it sound like an order, but an optional one. Just rediscovering the river might be done by his taking up the number two position again. But by daring a little more, the finding of his own course and of ours behind him, he would regather himself. I know — I remember — how a capsize starts you doubting your perceptions and reactions and how, just after you've had one, every little wave seems capable of filling or upsetting your boat and every rock and hole capable of lunging from yards away to seize and devour it. A few miles of unspoiled running are needed to start bringing back confidence. When it has returned to a certain level, joy begins returning, and you begin mixing yourself a fresh batch of that heady mixture that is one of the reasons men run rivers.

Having Don Ross run the lead was intended to tell him something, too. It was to say that what had happened in Satan's Gut had nothing to do with ability, or trust in that ability. He had simply been the protagonist in a drama that would have suited the Greek tragedians; his actions to save his boat from a dreaded fate proved to be the very actions that undid him, yet he could no more see ahead into the rapid than a Greek hero could see ahead in time or space. Perhaps fate was waiting at Satan's Gut that morning. Irony certainly was: the rapid's other name, given by Powell, is Ross Falls.

We pushed off shortly before noon, the *Bonnie Anne* ahead, the *Norm* back in her place, and the others in the same order we had been using. Running closely behind Don brought back a feeling I'd forgotten, that of following someone down a river — a good feeling, better than it used to be. I let my mind go back to the first trip through Grand Canyon, running second to Frank Wright, whom I thought then to be good but knew later to be the Best.

Those miles, following Frank and beginning to learn River were wondrous ones for a country boy who hadn't until then known that water could be read; that there was a technique for putting rowboats down through waves cresting two stories high. Comparing the Me behind Frank thirteen years before with the Me behind Don in Cataract Canyon, I noticed that both younger and older Me preferred to run the lead boat, but now I felt it a responsibility, as well as the measuring of myself against the river.

Don led us down through five miles of wonderfully sassy rapids, twos, threes, fours, and fives set close together like those of upper Cataract, speed-reading them nicely by standing momentarily at the oars before plunging in. After almost an hour of sustained, intricate running, I concluded he had his relationship with the river headed back into an untroubled perspective, and we made a late lunch. Two miles farther we reconnoitered Gypsum Canyon Rapid and ran it; two miles beyond Gypsum came the realization that it was not a rapid ahead ponding the current: the dead water was the head of Lake Powell. We checked the sheets; we were 3565 feet above sea level, at milepoint 194. Before the lake invaded its last miles the rapids of Cataract Canyon had extended downstream to Mile 177, dropping one hundred feet more in those seventeen miles. When Lake Powell is filled, dead water will reach even farther into Cataract, to cover the little beach where we had righted the *Bonnie Anne* at the foot of Satan's Gut.

Win took the oars then, so that Don could lie on the stern deck and close his eyes, and it seemed a natural thing to have him rowing alongside the other oarsmen. Both being rather solitary, Win and I have trouble keeping a conversation going, but I think of him more as crew than passenger and the river has given us a kind of kinship: once, a few years before, we had been capsized together in the *Norm* because I thought Granite Falls was runnable at that stage of water and he was willing to be the extra weight needed on the stern to find out. The object was to save a dangerous, boat-gouging lining job — which then had to be done anyway, after the *Norm* ran over an explosion wave and it blew up underneath and rolled us.

We had reached the head of Lake Powell in midafternoon. By three o'clock the next afternoon, stopping only long enough to camp at dark and resume rowing early, we had pried the seven boats down the lake to Mile 171 and were passing under the high steel bridge built there four years earlier to provide a highway crossing of the upper lake. The bush pilot who had landed at a nearby dirt airstrip to pick up one of our passengers stood on the bridge above, watching us slowly make our way. Our strokes after twenty-three miles of rowing were now yielding much less than ten feet each, and the two miles still ahead seemed like twenty.

"Hey, how did the Apollo team do?" someone from a boat called toward the bridge.

"They did it," answered the tiny figure above. "They walked on the moon."

FLAGSTAFF: FIVE DAYS

A few years ago I adapted a story to tell audiences what a man sometimes feels commercial river-running is doing to him. As it goes, three new arrivals at a nursing home were getting acquainted from adjoining wheelchairs on the sun porch. All three were withered and feeble with voices, like their owners, nearly used up and barely audible.

"Name's Smith," said the first. "Insurance business. Always lived a good clean life, yessirree. Never smoked, never drank. Early to bed and early to rise. Never been sick a day in my life. Be ninety-two in March."

"My name's Wilson," said the second, after a while. "Been in a lot of different things. Been a lot of places, too. Always did like to kick up my heels a little bit. Wine, women, and song, that's the story of my life. Never had any need for doctors or medicine either, and I'm ninety-four."

It was silent for a long time before the third feeble little man spoke. He was easily the most battered and worn of the lot, with a vacant stare in his eyes and hair like a bleached rag mop. His shaky hands hung over the ends of the chair arms partly closed, as if grasping something. "Staveley," he rasped weakly. "River guide since I was twenty-five."

"That doesn't sound so bad," observed Smith, with Wilson nodding agreement. "At least you've still got your hair."

"Yes, I know," said Staveley, "but dammit, I'm only thirty!"

The story was funnier to me when I *was* thirty. But I would go on to make the point that it's not the river — the wild water — that one associates gray hairs with; it's the logistical aspects of running it as a scheduled, guided adventure for others. I believe if a man could live his life beside his river, run it only for himself and only when he wanted to, on his own schedule (or *non*-schedule), he could easily outlive his contemporaries. There are many times when trying this is very tempting. But he can't, when he is young at least, just find an Echo Park somewhere and let twentieth-century life fly by at Mach Two. He needs the mainstream as much as the escape from it. And so a professional river runner is simply seeking the best of both worlds, adventure and livelihood.

But the methodical mustering of passengers, oarsmen,

boats, provisions, and equipment at a certain prear-
ranged date is the more exasperating of the least enjoyable of the two worlds. And during the time a river trip
is being assembled or disassembled, a river runner's
home is a tense and bustling operations center; a mixture of combat command post, priority freight dock, and
Grand Central Station. For five days in late July our
home in Flagstaff, Arizona, once again became such a
center, as it does just before the start of each run
through Grand Canyon.

We had reached Hite on Monday, July twenty-first, on
schedule. There we loaded the boats and equipment on
trucks and trailers to move by road around Lake Powell
to Lees Ferry, Arizona, where we would start through
Grand Canyon the following Sunday. By prearrange-
ment, several of our passengers' cars had been driven
from Green River to Hite to be available upon their
arrival. This allowed them to go their way unencum-
bered by our convoy of equipment. Rich rode to Denver
with the McKennas, Win returned the three drivers to
Green River on his way through, Yosh flew out, and
John and Mary Womer were met by Kent Frost, a well-
known jeep tour operator, to begin a back-country tour
of the Canyonlands, just east of Cataract Canyon.

The boat convoy's planned overnight stop was in the
little town of Mexican Hat, Utah, where my Mexican
Hat Expeditions were originated in the 1930's by Norman Nevills. We arrived just before dark, having traveled slowly on the ninety miles of back road, most of it
unsurfaced and rough, and found a long cool shower
even more welcome than the novelty of dinner eaten at
a table and cushioned chairs. Tuesday morning we resumed our journey, driving along the western edge of
Monument Valley, then angling across the Navajo In-

dian Reservation to Flagstaff. There the *Sandra* and *Bonnie Anne,* having been allocated to a trailer of their own, were unloaded so their crude field repairs could be supplanted by my father's craftsmanship.

One of the first priorities was obtaining an oar to replace the spare lost in Cataract. My own oars were now all in use but fortunately the owner of the *Bonnie Anne* had not sent her oars with her and still had them available. I immediately phoned him and he generously offered to put them Air Express on that night's Frontier flight from Colorado. Next, as the life jackets came off the load, we determined how many had been used as seat cushions and crushed — a curious but human trait, I suppose, to take a device that was carefully made buoyant for saving one's life and to destroy that buoyancy for the sake of comfort. On perhaps half the passenger jackets at least one of the vinyl envelopes containing the kapok fiber had been ruptured and the contents waterlogged. They had truly become what the boatmen sometimes call them jokingly: "life extinguishers." In such condition they could have been exactly that. A call to California started new jackets and buoyant inserts on the way, also by Air Express.

By Wednesday morning the floor of our double garage had been set aside as the staging area, which meant absolutely nothing was to be placed there except equipment and provisions for the trip, and nothing to be taken from it. Each item placed there is checked off a master list and it stays there until it's ready to load. I'm not entirely certain I'd even check bailing buckets back out of the staging area if the house was burning. Once you've started down a river canyon, there's no such thing as "almost enough" bedrolls, or life jackets, or screws, or as "almost" having brought a saw. One has just what one has left home with.

As everything was sorted out, the jobs began to present themselves: New plywood patch pieces / rewash and count the cups, plates, and utensils / get more tar compound / replace the worn cotton rope wrappings on some tholepin rings / buy a cooking pot to replace one carelessly lost in the river while scouring it / find another bailing bucket for the *Bonnie Anne* / buy nails and screws for the next patch job / call the airport to check on the oars / ditto for life jackets in a day or so. . . . New chores kept presenting themselves and soon I had three lists headed:

THINGS TO FIND   THINGS TO BUY   THINGS TO FIX OR DO

and a growing column of items under each heading. Each of the oarsmen had also left a short list, at my request, of minor repairs needed to his boat — a hatch seal recemented, a safety snap replaced or some such small detail that could be annoying, or, even worse, could lead to more critical problems. Those were incorporated into the third list. In addition to those lists, there were separate grocery and meat shopping lists, which I was saving until the day before departure.

By Friday the *Sandra* had nearly regained her smooth lines, the raw sandwich patch having been replaced by a carefully fitted insert of new wood, faired in with a putty of sawdust and fiberglass, and covered over with several layers of fiberglass cloth. When the injured areas had been repainted they weren't detectable from the rest of the hull. The *Bonnie Anne* was fitted with a new bow splashboard and spare oar clip, her stern splashboard straightened better than we had been able to do on the

Don and Dan survey some of the supplies and gear being readied for loading at Lees Ferry.

river, then touched up with new paint. By Friday evening, Fred and Maggie were back from the South Rim, Doug and Dan from Phoenix, Don from Colorado with permission from his doctor, and one in Flagstaff, to continue the trip. Chuck phoned from Page to say he'd meet us at Lees Ferry, and Rich's replacement, Walt Prevost, from Seattle to say he'd be on the 7:00 A.M. Frontier flight from Salt Lake City. That day there were finally more cross-outs than clean lines on all my lists except those for the groceries and the meats, and logistical matters seemed well in hand.

Saturday morning I took the grocery list and spent three hours loading a long line of basket carts from supermarket shelves, rather happy to have even that kind of respite from telephones, In and Out baskets, and the other logistical commotion. About noon the truck was loaded and ready to go. Fred and Maggie took it and a trailer carrying three of the boats, my father took a truck-trailer load of the other four, and they started north to Lees Ferry in a heavy rain. A few hours later, when every item on every list had been crossed out, Joan and Walt and I gathered up the few remaining pieces from the garage floor staging area, loaded them into the trunk of the car with several cases of dry-iced meat, and followed the others north.

By Sunday morning the lower end of the Lees Ferry launching ramp and an area at one side had become our new staging area. The seven boats rode the water lightly nearby, ready to receive their loads and, as it always does, the likelihood of getting into them everything that was on the beach looked slight.

The cans and bags and boxes of edibles were divided into as many equal piles as there were boats, and separated in terms of whole meals. One boat, for example, carried canned fruit for three meals, canned vegetables for four, bagged cookies for five lunches, eggs for three breakfasts, and so on. In this way each boat is fairly self-sufficient, but more important, the loss of it, which must be accepted as a possibility, will not leave the group deprived totally of any one item. As a precaution against confusion that may result if can labels are soaked off by leaks such as the *Sandra* received in Cataract, each can is code marked with a waterproof ink before being taken aboard. This is a chore Maggie always graciously undertakes, and of which she cannot fully know my appreciation. That she is able to get her own gear ready, in addition to labeling four hundred cans with "Apc" for apricots and "Gbn" for green beans and so on, is a tribute to her organizing ability.

Streaming down the Lees Ferry beach in seven columns, waiting to be labeled, were the cans and jars and boxes and packets of peaches, pears, apricots, grapefruit, cherries, plums, applesauce, beans, corn, stewed tomatoes, peas, spinach, yams, stew, spaghetti, ham, soup, peanut butter, jam, honey, salad dressing, syrup, pickles, toilet paper, coffee, tea, sugar, flour, salt, cereal, and dry milk, along with thirty-six loaves of bread, twenty dozen eggs, and fifty pounds of cookies. Missing because they were in cold storage at a nearby lodge were steaks, chicken, hamburger, bacon, lettuce, tomatoes, cabbage, and butter.

And at the edge of the launching ramp, in the shade of some sparse tamarisk bushes, our passengers sat with their duffel, wondering no doubt whether there would be room for them to go along, after all.

SUNDAY'S WATER

Lees Ferry is an historic starting and supply point for Colorado River expeditions. It is one of the few places the canyon-locked river emerges from its deep narrow gorge to allow boats and provisions to be hauled to its bank. For about a mile, just long enough to collect the little Paria as a sometimes flowing tributary, the Colorado makes a bend or two through lower, open country at the end of Glen Canyon, then heads purposefully downstream into rising sandstones and limestones to begin its Marble Canyon. Colorado River Mile 0 is at Lees Ferry; mileages are counted upstream to the confluence above Cataract Canyon and downstream to the Gulf of Calfornia.

Glen Canyon Dam is upstream at Mile 15. Before dam construction was started, Glen Canyon trips used to end at Lees Ferry, and Grand Canyon trips start, from a great umbrella-shaped willow on the north bank. John Lee's cable-tethered ferry of the latter 1800's was put out of business by a steel highway bridge about 1928 and his little buildings of stacked slabstone abandoned. Most of them are still there, historic relics. The clear, cold river below the dam has now been stocked with trout, and Lees Ferry is a fishing camp. The new, wide launching ramp is nowhere near the venerable old willow.

The water flowing by is now liquid emerald, but it moves too languidly to catch diamonds of sunlight. Its surface is lifeless, almost like a picture of water, rather than water. From habit one refers to it as "the river" but it is not any more, it is "waste water" from Glen Canyon Dam. The river was a feverish, silt-laden Presence that through late spring and the summer swelled to flow over the toes of our willow tree and urgently on into Marble Canyon. The waste-water river doesn't do that; there's not enough of it and all its spirit seems to have been absorbed by its plunge against the vanes of the dam's hydropower generators. It even permits slimy green moss to grow on its rocks, something the rushing, rasping old Colorado would never have allowed.

Boatmen discussing a river's flow talk in terms of "cuesecs" or sometimes "cfs," the number of cubic feet per second of water passing a bank-to-bank line imagined crossing that river. There are ways of obtaining this figure by measurements, which the Geological Survey takes at Lees Ferry and other fixed locations every day, but an oarsman takes pride in being able to "eyeball" it from shore with a few moments' study. It's part of his expertise, his stock-in-trade.

In the spring before the snow-swollen runoff was captured by Lake Powell, it used to charge past Lees Ferry at forty thousand, fifty thousand, eighty thousand cuesecs, for several weeks. In the summer of 1957 it peaked at one hundred twenty-six thousand cuesecs and ran a hundred thousand for three months. The spring runoff was always the river's time of housecleaning and renewal. When it began to subside, spacious new sandbars had been laid up by shoreline eddies and great rafts of driftwood had been snagged behind trees and bushes and on rocky promontories. We used to consider thirty thousand cuesecs ideal boating water, and twenty-five thousand minimal, because that stage began exposing rocks in the rapids. It would probably have been impossible to convince us we could run on six thousand to twenty thousand cuesecs of waste-water river. It still is.

Actually, the people of the American Southwest determine the number of cuesecs we can have to run the Colorado through Marble and Grand Canyons. The flow into the generator penstocks of Glen Canyon Dam is regulated by the same computer that regulates those of Flaming Gorge Dam. This computer is connected to a vast power-line intertie that reaches into the teeming metropolitan centers of Arizona and California. Electrical power demands from those centers are sent to the computer and it puts additional generators immediately on the line. Each added one sends another three thousand cuesecs or so down the canyon past Lees Ferry.

The complication to river runners lies in human life patterns. In terms of electrical demand, Southwesterners, like Americans in general, do most of their living from midafternoon until the early hours of the morning.

Thus there is a heavy demand for electricity from, say, three in the afternoon until three the next morning, then less demand from three in the morning until three the next afternoon. The computer reacts to these two different demands by letting a larger volume of water through the generators for twelve hours, then a smaller volume for the next twelve. The effect of this is to send higher water down the canyon for a half day, then lower for the next half day. By the time the low water begins, the front of the high water has traveled about forty miles downstream, making a forty-mile-long plateau. The lower water follows it as a lower plateau about the same length. A simplified cross section of these alternating releases would look like the up-and-down tops of a column of tables of two different heights pushed end to end, a high one then a low one then a high one then a low one and so on, all moving downstream. When you get on the river at Lees Ferry you can either get on a high table or a low one, depending on which is going by at that time. Or you can wait.

That is the weekday pattern. On the weekends in summer, however, the people whose light switches and television sets regulate our water spend most of their time outdoors and out of the city. They begin their exodus on Saturday and it becomes nearly total by Sunday. When ten million people stop talking to the computer through their electrical circuits, the computer begins sending us Sunday's water.

Sunday's water is the worst aspect of the regulated river. Like weekday water it has the high–low pattern, but at a flow so reduced that the levels are more in the nature of low–*very* low. With the top four or five feet of depth taken from it, the river begins to narrow and slow, and show many, many rocks. Running it becomes much more dangerous.

Nevertheless, I had scheduled a Sunday departure from Lees Ferry because the fourteen-day trip would have to contend with the very low water somewhere along the way, and I would rather we did it there. With that schedule we would be on it about a day. Our Monday night camp would let the Monday water begin catching up with us. After that we'd in effect be running ahead of the next Sunday's water, so that by the time it caught us we'd be nine days down the canyon and only have to cope with it once more; then we'd be off the river before it arrived again. Especially important was the fact that those nine days would put us beyond the Upper Granite Gorge, where extremely low water is an even more serious and dangerous problem than it is in Marble Canyon.

The river was now rising about two inches an hour but was still very low as we began moving everything down from the stacks on the beach and fitting it into the boats. The previous higher flow had left several yards of slippery mud at the lower end of the gradually tapered ramp, and a slimy edge on the steeper shore adjoining. The boats floated in five or six inches of phlegmy green water through which showed moss-bearded rocks and patches of sunlit sandy bottom. Our passengers still sat in tamarisk shade watching seven oarsmen and three or four others working unhurriedly with gear and provisions, and not really understanding our need to wait for higher water.

The places of Rukie and Keri Jelks and their sons, and of Win, and George Philips, were being taken by seven new passengers for the next segment of the expedition. Charlie Philips's pretty young wife Rosemary joined

him to take his father's place. The two younger Mc-
Kenna children, Katherine and Andrew, joined their
brother Douglas, who had been with us from Jensen.
Judy Rooney and Tommy Dunlavey of Senator Barry
Goldwater's staff filled two more of the seats. Completing the roster were two young men I was perhaps understandably proud to have along, our sons Cameron, thirteen, and Scott, ten. For all of their young lives they'd
been seeing movies and slides of the trips, and waving
good-bye to me on one riverbank and meeting me days
or weeks later on another. For a long time it was impossible to explain to them how the same river can be more
than one place and that the river I came off was the same
one I'd gotten on upstream. As they grew a little older,
we tried some short trips on the lively but mild San
Juan River at Mexican Hat. One of the boys, when he
was four, reacted to the bobbing of the boat and slapping
of waves against the hull by crawling back under the
stern deck and crying himself to sleep. A few years later
they were taking turns sitting in front of me on the seat,
trying to reach around the big oar handles while I let
their hands follow my pushes and pulls. Then they grad-
uated to "innertubing" the San Juan's riffles, running
"stern first" behind the boat with their hands as paddles.
At thirteen and ten they were ready for big river, and I
wanted them to see the inner Grand Canyon before it
changed any more.

By about two o'clock Sunday afternoon, enough air
conditioners had been turned on in the hot Southwest
that the computer was sending another foot of water
past Lees Ferry, enough to start down the river. Duffel
was stowed, the frozen meat was brought down from
cold storage, and everyone chose seats. Most of the continuing passengers elected to remain with the boat and

oarsman they had ridden with through Cataract Canyon and the new ones filled the remaining seats. We said our good-byes and rowed out into the slow green river, forming our column of seven.

As we drifted down through the long, wide riffle at the mouth of the Paria, I again caught myself looking through the shallow water at the river bottom instead of concentrating on the surface of the water. To the eye trained to plot a safe course by reading the top of the water, a clarity that lets it see through to the bottom is not only distracting but dangerous. Rather than the wave patterns they made above, my attention was caught by the bright red and yellow stones on the river bottom four feet under the boat, and by the surprising speed we made over them, much greater speed than the passing shoreline conceded. Then gradually the river narrowed to begin a little rapid below the Paria, and the bottom dropped out of sight. Dregs of a storm-water runoff were coming down the Paria into the Colorado so that its right half was silty gray brown and its left half sparkling green, entering the rapid. The waves were mixing the two waters so that they rolled out at the bottom more nearly the color of the Paria, and much easier to read. Beyond the rapid, the yellowish gray Kaibab Limestone began rising gradually close against its banks, walling off our view of the reddish sandstones and red-gray-purple shales that form the Echo Cliffs and Vermilion Cliffs between which Lees Ferry is pocketed. This up-slant of the land, combined with the down-gradient of the river produces within very short distance a canyoning effect that continues to deepen noticeably for more than sixty miles. At that distance downstream, as one regards the rim of Cape Solitude above the boat, one is looking almost four thousand feet straight up.

199

Almost an hour after pushing off, we came within sight of the highway bridge crossing the canyon, which had already deepened more than four hundred and fifty feet. Somehow it has become traditional for those who watch us shove off to hurry down from Lees Ferry to drop huge slabs of flat rock from the bridge as the boats approach, and soon we could see the first slab on its way. It came down as if in slow motion, getting larger, turning casually, seeming to float like a leaf, then disappeared into a tower of leaping water followed in a second or two by its *WHOCK!* as the sound shot upstream to us. The cars that occasionally crossed the bridge were nothing more than dark mites high up against the gathering clouds, and the individual figures of our well-wishers only tiny tally marks. The bridge, and its traffic and its people splatting rocks into the river ahead of us, always seems more of a threshold to the Grand Canyon adventure than does the moment we take to the river at Lees Ferry, probably because the bridge is a concise silver line drawn across the canyon, and we don't see anyone on the rims after that.

Badger Creek at Mile 7.8 is the first major rapid of Marble Canyon and is a topographical rarity in that two side canyons enter the Colorado directly opposite each other. Both have dumped in the storm-water boulders that cause the rapid, but it is named for the one that enters from the right. The effect of two boulder fans being thrust out to meet each other in mid-river is to create a dike of rocks from bank to bank, over which the river must find its way. When there is enough water, it does this by forming a tongue that has thrust away some of the boulders about midstream, but when the water is lower it lies behind the rocks only able to sluice between them as it would between the nubs of worn old teeth.

Navajo Bridge — silver steel threshold to the Grand
Canyon adventure

It was for Badger Creek we had been waiting for the river to rise. The only way through it safely is down the midstream tongue. But the drop is a rather instant twelve or thirteen feet, making the proper point of entry impossible to see until the last three or four boatlengths from the lip of the dike. At that point one had better be in position or one will find oneself missing the tongue and plunging in between two of the rocky teeth. For some reason of perception or parallax, misjudged approaches to Badger Creek are almost always made too far to the right rather than the left. Perhaps this is because, lacking any surface features of the river to use as markers, one hedges and takes the exact midstream to be able to move either way; then in the instant one has for orientation, discovers the tongue is a few yards left of middle.

It has always been my practice to land above Badger before running it, regardless of whether I believe I've learned to judge just where below the brink the tongue lies. The stop is as much for psychological reasons as it is a navigational precaution. To go booming over the brink of Grand Canyon's first major rapid with freshman and sophomore oarsmen following is to infer that it can be run in an arbitrary way over a course that is not too important. It is impossible to put oarsmen in awe of the major rapids unless one starts with the first, and to fail to put them in awe, or to try to, is to do them a disservice that could have damaging or sorrowful consequences. A properly disciplined oarsman takes the attitude that the rapid he is about to run is the worst one on the river, whether it is or not, and thinks of its structure, and his course and maneuvers, with almost surgical precision.

We tied along the left bank, well within the canyon-filling roar, and made our way downstream on the coarse, steeply pitched slide-rock, at the same time angling upward in order to reach a vantage point. Opposite the head of the rapid and about seventy-five feet above, we found a great block of tumbled limestone on which seven oarsmen and three or four interested passengers could stand to study the pattern of rocks and water.

Badger is a straight rapid with a crooked tongue. The water runs quite directly through it because it is deflected equally from each side by the rocks of the two opposing canyons. But the smooth tongue itself, after pouring over the brink, begins a turn to the left that becomes almost a lateral hook. Probably there is an immense and very deeply submerged rock at the point of the tongue, turning the water to the side and simultaneously banking it steeply to form the upstream slope of the first tailwave. Water surmounting this wave goes on to form other tailwaves below; water hooking left out of the tongue moils and then runs down over some left-side shoals. At higher water, on the natural river, the tongue was wide, and thrusty enough to straighten its hook, and was runnable down the middle. Now the course is a touchy one down the right edge of the tongue, past its bent tip and into fifty yards of tailwaves that run out below.

We spent a half hour studying the rapid, mostly the tongue. I let the motion of the water carry my eyes down it over and over, as if they were floating there, feeling the water thrust them along as it would the *Norm*. Nuances of current pulled and pushed them as they traveled the right side of the sleek water, until they found the point at the top from which they were carried through without being drawn left at the bottom. Then we discussed this entry in terms of boatwidths from the nearest tooth of rock. Halfway down the tongue, perhaps two boat-lengths, a little white-tipped lateral wave rolled in onto the smooth water; that, we decided, was to be our marker for sliding down the tip past the hook. Because I was still bothered by the Satan's Gut incident and the thought that signaling the boats into it might have saved a capsize, I decided to divide the boats into two shorter columns for the run. Fred would signal my group in, then when we'd landed below I'd return to the talus and signal him where to lead the others in. Perhaps this gave some of the oarsmen a sense of security that they shouldn't have had; I sensed some lack of concern among them; I tried to counter it by prolonging our stand on the rock and by reiterating the idea that the run would be blind until the very brink, then necessarily very precise, over the little lateral wave and off the smooth tip into the stacks below.

Fred stayed on the rock, along with several others who would be in his group and wanted to watch the first run. Returning to the boats, we announced that all passengers could ride, and that it would be a two-part run, with four boats completing it before the last three began. Cameras were stowed, life jackets tightened, and hatch tighteners snugged down. When we had moved out to mid-river as an initial position I could see Fred high on the left downstream facing me. As soon as his eyes could project the line of the current that was carrying the *Norm* he raised his left arm and held it. When I had moved slightly left, he raised both arms over his head to say "hold" and the boats behind me moved accordingly to come down the same course. Then at the brink I could see the sleek brown tongue just beyond the stern and the little wave halfway down it. In a few seconds the mound at the hooked tip was going by, and we rode over a dozen tailwaves and caught the edge of the left-bank eddy that carried us back upstream on its reverse current. It was all over in perhaps a half minute. Walking upstream across a big sandbar, I crossed the rocky bed of the side canyon and climbed to take Fred's place on the rock. In a few minutes the *Camscott* came into view in mid-river, and with a minimum of correction, slid down the tongue, followed by two other boats. It was not a very encouraging start; of the seven boats there had been three bad runs. In two of them the boats had somehow been allowed to swap ends and had run the tailwaves bow-on, Powell fashion. The third one had missed the tongue entirely, gone down a chute between two rocks, miraculously missed big boulders below the right half of the dike, and come out with a great deal of water in the cockpit. On board that one was Scott, my ten-year-old son.

We had planned a short day to let some of the low Sunday water go on by and the higher weekday water begin catching up with us. As we went into camp at the foot of the rapid the river was still rising, and several times during the evening we had to take up slack by twisting sheepshanks into the bow lines or retying them altogether.

The night showed some promise of being another like the one at the confluence of the rivers. Throughout the afternoon a storm had been mustering to the north. At Lees Ferry we had been able to look west across the more open country to see clouds hanging over the Vermilion Cliffs and the high Kaibab upland beyond. Then as we started down the river with our minds on what lay ahead, and the deepening of the canyon closed us off from the outside world, we forgot about the sky except for the deep blue strip visible directly overhead. At

Badger Creek the canyon notched the limestone horizon deeply, revealing that those same clouds had grown and darkened in two hours and had moved toward us. As we studied the rapid, they were advancing to form a low gray roof over the side canyon, and by the time we had finished the run they had edged out the late afternoon sun. With its warmth preempted by them we found our wetness from Badger's waves almost chilling, even on that July afternoon.

As dinner was cooking the wind began to rise and pick up the dry beach sand so that the pots of salad and soup and vegetables had to remain covered between servings and the steaks had to be hurried across the grill. A number of passengers and boatmen finished their meal quickly and strode off to make up rain beds before the skies opened. A few sporadic drops were already falling.

My sons have slept outdoors many times, but not in wind and blowing sand, and I helped them arrange their groundsheets so that the wind tended to blow them closed rather than open, and to weight the edges with heavy pieces of log and with rocks. With an extra plastic groundsheet, I wrapped up enough firewood for the breakfast fire, and then took an armload of tube tents and went to see how well waterproofed the passengers were. Most of them were already wrapped in their plastic cocoons ready for what might come, but a few yards farther I heard voices coming from three cone-shaped mounds just visible in the darkness against the lighter sand. They proved to be Judy, Tommy, and Chuck sitting cross-legged, draped in ponchos. They were doing fine, they said, just waiting for it to rain or not to, and they invited me to join them for a nightcap.

"You probably won't believe this," Tommy said, "but this is the first night I've ever slept outside in all my

life. Not bad for a starter, eh, a sand blizzard on the
Colorado River?"

He was doing well, though, as he had said, and had
inflated his air mattress and weighted his groundsheet
with rocks to keep it in place. I showed him the plastic
tube tent I'd brought him, and how his bedroll could
be slipped inside as into an envelope, and made up a
similar one for Judy. Chuck, remembering his night
standing in the dripping bushes at the confluence, had
dragged together a framework of driftwood logs and
roots and draped them with plastic to fabricate a prime-
val den that would have gladdened the heart of a hiber-
nating bear.

The four of us sat in the intermittent drops and talked
for perhaps an hour; then the rain began falling more
steadily. Tommy slid down into his tube tent, Judy into
hers, and Chuck disappeared into the crawl-hole of his
den. I picked my way past a dozen plastic-wrapped
bodies, back down to the river, and found my bed pit
beside the bow line of the *Norm*. The rain lasted for
about two hours, never quite becoming a downpour,
though the sand continued to blow for several hours
more. Then the morning sun rose into a clear warm sky.
Everyone reported a reasonably dry and restful night,
and the breakfast coffee seemed especially good.

My plan for the day was to go as far as Mile 30, because after Badger Creek, campsites are rather scarce
for twenty miles. The consistent up-slanting of the land
soon brings the wall-forming Supai Sandstone and then
the Redwall Limestone to river level, and the Colorado
has been able to grind only a much narrower, vertically
walled gorge through those strata. Most of upper Marble
Canyon's shoreline consists of cliffs or ledges rising directly from the water; sandbars of camping size don't

begin to appear until less resistant strata have been lifted to the cutting level of the river.

The agenda for the day was twelve Marble Canyon major rapids:

MILE	RAPID	DROP
11.1	Soap Creek	16 feet
14.4	Sheer Wall	9 feet
17.0	House Rock	11 feet
20.6	North Canyon	11 feet
21.2	Twenty-one Mile	12 feet
24.1	Twenty-four Mile	4 feet
24.5	Tanner Wash	9 feet
24.9	Twenty-five Mile	6 feet
25.5	Cave Springs	6 feet
25.8	Twenty-six Mile	4 feet
26.6	Twenty-seven Mile	7 feet
29.2	Twenty-nine Mile	6 feet

Our boats were high and dry now, the previous afternoon's small rise having moved on downriver through the night. It had been followed by the low of the early morning generating cycle, which was still passing but beginning to increase now that the people of Los Angeles had plugged in toasters and coffeepots. We untied the bow lines, still taut from having been taken up to prevent the boats chafing each other on higher water, and dragged each of the seven down the sand to the edge of the water. It was hard, tedious work that the so-called "taming" of the river has made unavoidable at certain points downstream from the dam.

The morning was pleasantly still, as we drifted down to Soap Creek on about seven thousand cuesecs. A thousand feet of height had been achieved by the walls of the gorge, and they were reaching steadily for more. Much of the river was still in shadow, and the air had not yet been loaded with the beginning of the midday heat. There was hardly any sound except the sweet, amusing trill of the canyon wren that sounds like some kind of whistling toy in need of rewinding. Once the children were able to watch a wary fox sneak away among the talus boulders, and a little farther to watch a huge beaver plow a vee of wavelets across in front of us as he hurried to hide among shoreline rocks and roots.

Within an hour we had reached the head of the rapid and were crowded onto the top of an enormous limestone block, reconnoitering. The canyon turned rather abruptly south just above, so that most of the rough water was still in the shadow of the eastern cliff. It wasn't until we'd spent quite some time on the rock that the sun-shadow line backed across the river to let contrasting light fall on the troughs and crests, so that we could be certain distances were as they had appeared to be in the shade.

The success of the run depended on a steep, pointed wave we called the "tent." It lay just beyond the end of a rather scrappy tongue that in higher water would have smoothed and pushed down into the rapid. The "tent" stood about six or seven feet high and was the beginning of the tailwaves, which were longer waves of about the same height crossing most of the river below. Toward the right bank, near us, the crests of the waves became more ragged, tending to fall back upstream into their troughs. The smoothest course lay over the right shoulder of the "tent" and stern-first through the remainder of the rapid. We picked out a small lateral wave on the tongue as an entry marker, and an eddy line that would lead us down to it, and returned to begin the run.

As we loaded, one of the passengers, concerned about Scott's rather wild ride through Badger, drew me aside and insisted on trading seats so Scott could ride in the *Norm.* It was a generous and even a brave gesture, yet

I knew as I accepted that riding with me in the proof-boat was no guarantee of safe passage either. The only justification I could give myself was that an adult could probably take better care of himself in wild water in a life jacket than could a ten-year-old boy, if the same oarsman got two bad runs in a row. I calculated no upsets, but quite a lot of water on board all boats from the ragged waves.

I was wrong. Scott's previous oarsman made one of the better runs, and the *Bright Angel* was capsized somehow, either by broaching on the "tent" or in a trough just below it. Looking back when the tailwaves had subsided I was aghast to see a wrong shape, the curved unpainted bottom of the boat, coming down through the rapid, with Rosemary, Charlie, and Dan clinging to the safety line along the side. Anger and disgust flared inside me momentarily, as I pulled with the *Norm*'s bow cutting the current, making it carry them to me. When the *Bright Angel* touched we swung it in below; then, by its bow line, we let it into an eddy on the left. Bailing air under it as quickly as possible, we rolled it upright and opened the hatch covers, finding they had held reasonably well, leakage into the stern wetting the outsides of some film cartons but not damaging the film, and into the bow soaking two or three pounds of poorly packaged cookies into chocolate mush. Rosemary had lost her sunglasses, but no one had been injured. By the time the boat had been sponged out and repacked, it was nearly noon and lunch was spread on the rocks. It was hard to muster much appetite, but we would perhaps make up some time by eating now.

The river was still rising. When we pushed off, it had covered the small sand flat from which we had unloaded and reloaded the *Bright Angel,* an increase of three or four inches depth. The next few rapids, Sheer Wall, House Rock, and the beginning of Marble Canyon's "roaring twenties" were run as rapids should be; we approached them, read them from the cockpit, and plunged through, all getting good runs, successfully treading that thin line between making the day's miles and taking time to land and look at every rapid.

Then, entering Twenty-four Mile, the *Camscott* got slightly too far right. There was a short, sharp tongue into a half-dozen steep waves below, and I think I knew when she started down the middle of it she was going over — the fast drop into the trough behind the first wave had alarmed me and I'd skirted it to the left. She dipped out of sight. In a moment her stern appeared, trying to rise over the crest of the first wave, but sliding back. Then her whole left side rose into view and simply continued rising until her downside was carried underneath it and she rolled very smoothly and was over. She came out of the waves with Fred and Maggie and the McKennas clinging to her.

Once again the same disgusting drama was replayed, holding against the current, catching the boat, swinging it in below us, trying to get out of the current and into an eddy. I brought everyone aboard the *Norm* because I could handle the extra load better there than swinging from the *Camscott*'s bow line. Then loops of the line washed around my right oar like a coiled serpent, making the oar useless. I tried flailing the oar blade in an opposite circle to whip off the heavy rope but it hung too heavily in the current. One of the other boats moved in to try to free it and I heard myself bark to get the hell out of the way so I could work. By then it was too late to land the *Norm* and its load of seven and its dragging, pendulous burden above the next riffle. There was nothing to do but jump up and jerk the tholepin out of its pivot hole to free the oar and tip it to let the rope slide off the end and jab it back in place and start ripping at the water with both oars. The boats sloshed through together and I got them beyond an eddy line and against a little sandbar on the right bank. There wasn't much to say except, "Let's get her turned over."

She had taken water into both the bow and stern compartments, past seals designed to resist water over the deck but unable to cope with ten minutes of submergence. Everything in the stern compartment had been tumbled into an intertangle by the action of the rollover. Two or three jars of pickles and mayonnaise had been broken and had to be carefully removed. Cameras and duffel and bedrolls were soaked.

Now it became apparent the river was rising on our small sandbar, behind which there was only a coarse rockslide, and that our time there was limited. The day had been enough for everyone, and Tanner Wash, a problem rapid, lay waiting just a few hundred yards downstream. For reasons of morale, my own as well as the group's, I determined to make camp, provided we could all cross the current without being taken beyond the next point of rocky shore. On the opposite side was a high shelf of sand that would put us above rising water and at the base of it an eddy circled slowly back upstream. If we could cross the riffle in time to catch the toe of the eddy we could camp, dry out, and get psychologically reorganized. If anyone missed, we'd have to run more river.

The *Norm* tried it first. We were in quiet water beside the lower end of the riffle and to give ourselves all the distance we could, had to row up alongside and then stick the bows of the boats out into the fast water and angle across as quickly as possible. It proved to be a rather marginal procedure, with the river throwing the bows around strongly, so that most of the pulling had to be done with the downstream oar to stay headed across. But all seven succeeded in reaching the other side above the point where the eddy gave way to downstream current. As each came into the eddy, he rode its reverse current back up to the base of the sand shelf, and we finished directly opposite where we had started. Within a few minutes there was nothing to be seen of our little quay over there. Sunday's water had gone on down the river and was being followed by the weekday rise. Within two hours the bow lines had been taken up a number of times and the river had risen to cover a four-foot stick I'd jammed into the sand beside the *Norm*.

While it was still light enough to see, I helped Scott hollow out a sleeping-bag pit on the edge of the sandbank and scooped out my own, close by. The evening would give me time to recollect everything that had been pushed back because there were people in the river and boats to catch before they were swept away and wet cameras to get opened and drying. I pulled out the feelings of the day, and of the last few years, and examined them.

When each of the upsets had happened, I'd felt hot surges of anger, not at the oarsmen, but that such a thing could happen to them in spite of their skill. It wasn't always this way. Cycled waste water had made the river a difficult training ground. Before the dams, on the natural river, margins were greater. You could run a course to the nearest boatwidth instead of the nearest six inches and it would be a good run. An oarsman had a chance to learn, then, without the penalty of upset boats and distraught passengers. He could begin using the technique in gross form and then refine it with experience; the river gave him time and space to learn. Now he is pressured toward impossible precision from the first oar pull. Bright "naturals" like Doug or Dan can still be trained there, but while that's being done, things are going to happen and almost happen that wouldn't have on full, living river. Giving a freshman a boat and pair of oars now is something like putting a nondriver at the controls of a formula road racer and teaching him to drive while competing in the Grand Prix.

On the river the anger is a selfish one, at the way waste-water cycles complicate river running. To find each rapid at its best, one would need a complicated schedule geared to the safest stages of water. And even if one *could* arrange to run each of the rapids "by appointment," what is he supposed to do when he needs high water for one, but it happens to always go past that certain milepoint in the middle of the night?

On shore when the day is over, the anger becomes broader and more seething. It becomes anger at the mentality that could wrest living rivers from nature, turn them over to a thinking machine, and call that progress. It was the only Colorado River we had, and we had it for less than a hundred years. For a time I thought perhaps it had only been changed a little by the damming. They told us it had. And after all, there was water still running past Lees Ferry and down through Marble and Grand Canyons. But it took only a run or two to discover that waste water is not the River. The River didn't toady in its bed, a sickbed that it can't clean. The River didn't have forty-mile-long troughs and forty-mile-long crests of water sent down it every day; the River began swelling with meltwater from the mountains in April or May, peaked with virility in mid-June, then began a subsidence that lasted through August or September. The River poured a broad tongue of water down into Badger Creek, not a narrow hard-to-find sluice between the rocks. There was a fine campsite, one of the nicest, at the foot of Soap Creek; now sand is no longer laid up there, and what was there has drained and blown away. Crystal Rapid rated perhaps a "two" before a side canyon flashflood virtually dammed the river with a glut of boulders the waste water can't move; now Crystal is unrunnable with rowboats. Lava Falls was sixty feet wider before the same thing happened there. The wide flat sandbars down through the miles I call "the two hundreds" were wonderful places for children to run and play at camp time, until waste-water cycles inundated the sand regularly every day and those clean wide bars were invaded by phreatophyte thickets.

And what about the changes that are beyond an oarsman's understanding? Where can we get a straight answer? Is it true the river's rare squawfish are now near extinction? And why have the great blue herons moved out of the inner canyon since the river was computerized?

The computer becomes a hate-object, a representation of all the pork-barreling and slide-ruling and empire-preserving that can forgive itself the destruction of irreplaceable grandeur for seven years of jobs, a monstrous concrete showpiece that has begun to leak so badly it is hush-hush, a cost outlay that will probably never be returned, and a reasoning that can impound the river in a reservoir so that evaporation sucks millions of gallons into the dry desert sky — and then name that reservoir after Major Powell, the forerunner of water conservation.

I lay there on the sandbar now angry with myself, too, for my sentimental attachment to the little wooden boats that are so much fun to run and that let a man find out something of what he can do and what he can't. They were designed thirty years ago to carry passengers on the Colorado, a deep, powerful, living river. This isn't the Colorado any more. A man can run rowboats for himself, for fun, but it seems foolish to try carrying passengers in them now that the margins are slimmer and the risks greater.

"The pontooners are realistic," I thought to myself; "chain some tubes of air together, put a motor on the back, pile people on it, and drive it down the canyon. You don't have to know anything about reading water, just dump it in at the top and it comes out the bottom. Smother the waves and holes and slither over the rocks. Any fuzzy-faced kid can do it and a lot of them are. Such imitation river-running is appropriate to an imitation river." I knew, had known for some time, that when our long rowboat run was over I'd have to become a pontoon driver, too, to stay in the business of canyoneering. And then in a few more years, plugs of unmovable side-canyon boulders will probably have created a series of sheer falls unrunnable even by the "baloney boats," and it will be all over for everyone.

By the time darkness had come, I'd brooded myself into such a state that sleep seemed hopeless. Then, after a while, Scott stirred in his bedroll close by.

"Daddy?"

"Mmm?"

"I had fun today."

And I'm sure I went to sleep wondering whether it wasn't a mistake to be teaching my sons to love the Colorado River.

CONQUEST & CONTEMPLATION

We hadn't lined Tanner Wash Rapid in twelve years and the realization that we had better was a disappointing one. Not having had to line in Lodore or in Cataract, the two canyons where it was most expected, we had run everything for four hundred and eighty miles. I had expected to go another fifty miles, to Hance Rapid, before having to let the boats down along shore. Perhaps it was because we had always run Tanner on the higher water of afternoon: having stopped short of our Mile 30 goal, we had lost time and were now seeing the rapid on the low water of morning. Perhaps, too, it was partly reaction to the previous day's upsets, but it didn't look as though seven boats out of seven could make the run without mishap. Two violent low-water holes bordered the tongue, which curved so that the lower right one was almost below the upper left one. It seemed very much a case of "if the right one don't get ya, then the left one will."

When boats are lined they are left in the river and worked past the rapid in the chutes and shallows near shore. The shoreline current supplies the downstream thrust against which several oarsmen let the boat down slowly by its bow line, while one or two others help it over and through the rocks and narrows. A full-scale lining is nearly as dangerous, sometimes, as the run it proxies, and is harder on the boats. Fortunately, the Tanner Wash operation was a limited one, needed only to bypass the left-hand hole, beyond which the rest of the rapid could safely be run.

Above the hole along the left bank the slowly moving current allowed each boatman in turn to drift down without passengers aboard a few feet from shore, while a team of two others held his bow line and walked along with him. Sliding through a narrow chute between big rocks, he rowed into a cove of dead water below a little promontory and was held while he shipped the oars. The boat was pushed out around the next promontory then snubbed below it by the linemen, and when that had been repeated around a large rock a few yards farther down the shore, the boat was just inside the left-hand of the two dangerous holes. At that point the oarsman readied himself, the bow line was coiled and laid into the cockpit behind him, and he steered down past the edge of the hole, along the tailwaves, and landed below in a huge eddy against boulders to reboard his passengers. It was

The decision whether to run a rapid or line it, required of oarsmen at a half dozen of Grand Canyon's worst rapids, is best made from a vantage point along shore, where the positions of rocks and thrusts of current can be studied. Left to right: Walt Prevost, Don Ross, Gaylord Staveley, Doug Reiner, Cameron Staveley, Fred Eiseman, Dan Gernand, Chuck Reiff.

an easy and almost riskless job and we finished by mid-morning. The full-scale linings ahead were to be more difficult and dangerous.

Tanner Wash, more often called Twenty-four-and-a-half Mile, is always a bugaboo lined or run, because one remembers it has claimed more lives than any other rapid in the hundred years man has been running the river. The structure seems simple enough but the positioning of the two holes is deceptive and one of them will almost certainly be taken in trying to avoid the other. Beyond those the tailwaves are ragged and sharp-troughed and much, much larger than they seem from shore; then, when it appears the rapid should have run its course, its tail pushes almost irresistibly into the left-hand limestone cliff. Two members of the Brown-Stanton railroad survey of 1889 were lost there when the river impinged their boat against the limestone and overturned it. Bert Loper, once "the Grand Old Man of the Colorado," was lost in the tailwaves in 1949, trying for one last run of the river to pre-celebrate his eightieth birthday. There have also been many "almosts" in that fraction of a mile.

After Twenty-four-and-a-half there are five more rapids that are moderately challenging, but even as they are being encountered the canyon has begun to change, to be more inviting. In the upper Twenties the Redwall Limestone begins to strike upward into view in the stratigraphic sequence. The Redwall is beautiful rock, a cliff former, a cave maker, a bearer of subterranean water to seeps and gushes. When Redwall becomes the shoreline, the shoreline is more friendly, even though it may rise straight from the water for hundreds of feet. When it has pitched on upward to take its place above in the many-layered canyon wall, it is still an attractive stratum, one that the eye catches at a glance, wending its uncluttered sheerness along the convolutions of the deepening canyon. Up there it is a warm creamy orange because eons of rain have sent an iron oxide "rinse" down over its face from red layers above. At river level, silt-scrubbed back to its true colors, it is the soft gray of fog or the near white of oystershell, or a horizontal banding of grays and whites. The old silty river, where it ran against walls or fallen blocks of Redwall, has faced them with sharp-edged flutes that look like the scallops spoons make in melon. One feels an urge to see great buildings and plazas of the prescalloped, prepolished gray blocks, or perhaps has the impression one has. It so resembles fine building stone that Powell named Marble Canyon for it.

In the Redwall such things as Vaseys Paradise can happen. Coming down a narrow, sheer-walled corridor of river, you look ahead a half mile to see a patch of astonishing greenery clinging to the limestone wall, and then widening lushly to mantle a rocky buttress that comes to the water. Veils of feathery white water cascade from holes in the Redwall, down through fingers of verdure reaching for them, then collect below in little rushes from which canteens can be filled. Poison ivy grows up among the greenery, but so do mint and watercress, and our stop at Vaseys yielded, as it nearly always does, some bonus ingredients for the evening salad.

We left Vaseys after a lunch made beside the cold, delicious water under a gray sky that was leaking a few drops of rain, but we rode easy river now. There were no major rapids, nor would there be for several miles, and the boats relaxed into a random flotilla, moving back into the file only for an occasional riffle. We passed the Bridge of Sighs, named for its resemblance to that historic Venetian portal to torture and death, but the youngsters and several others were napping and didn't notice. By late afternoon we had made a nineteen-mile day.

President Harding Rapid was our camp for the night. The "rapid" is nothing more than an immense block of Redwall Limestone that has bounced into mid-river to split the current and tear a rather impressively roaring hole in the water in its lee. It is believed the block fell sometime in the last hundred years. The unweathered, triangular cavity it occupied before breaking loose can be seen above, in the right-hand cliff. The rapid was named by the 1923 river survey party, who suspended their work and camped there for a day when they received word of President Harding's death by shortwave radio. The body of Peter Hansbrough, who was lost in Twenty-four-and-a-half, is also at Harding. Finding his remains there as they continued the railroad survey, the other members of the group buried him "with a shaft of pure marble for his headstone, seven hundred feet high with his name cut upon the base, and in honor of his memory we named a magnificent point opposite *Point Hansbrough.*"

The Harding camp was a pleasant one. Afternoon clouds had given up trying to drop rain on us and the day ended clear. We hadn't regained any of the second day's lost mileage but had made about what we should have the third. There was time before dinner for a canned cocktail, or to take pictures, or to make up bedrolls, or climb up through the mesquite trees that lined the beach, to visit the Hansbrough grave. Scott and Andrew played "King of the Hill" on the steep sand between the boats and camp, until they were sheathed in sand and it was all through their ears, hair, and swimming

suits. Cam and Douglas went down along shore to explore a low sandbar of several acres that had been
temporarily uncovered by the water cycle, their minia-
turization against the backdrop of downstream cliffs a
yardstick of the towering immensity the canyon had
achieved in only forty-three miles.

The water rose five feet overnight. We awoke to find
the boats nuzzling the sand not far below our bedrolls
and the older boys' sandbar completely covered by the
swelling of the big eddy. The rise, being the previous afternoon's increase from the dam, had been expected, but
was still gratifying. Its depth would be needed to run
Kwagunt and Unkar Rapids, and its somewhat greater
velocity helpful in making the day's run to Mile 75 so
that we would enter the Upper Granite Gorge on time,
in order to reach Bright Angel on schedule.

High above camp, the rim and the craggy points and
pinnacles just inside it had already been found by a
heavy sun and had become the hot reds and hot browns
of a desert summer day. Below the rim, the morning was
still soft, diffused. The canyon at Harding was in shadow,
but just below, by turning to the west, it opened itself
to some of the sun. Rays of sunlight thrust down across
an inner canyon haze that incandesced them into streaks
of almost eye-hurting brightness against the mellow
pastel background.

The river carried us along at five or six miles an hour,
riffling a little once or twice a mile, and after three
miles turned into a long, south-running corridor that
put it again in total shade, and the air was noticeably
cooler. Near Mile 50 a long, narrow bank of fog — perhaps a cloud hiding from the heat above the rim — hung
just over a low marshy section of shore. Two miles farther, we drifted through the shade line and out into the

sun for the day and crossed the northeast boundary of Grand Canyon National Park at the mouth of Little Nankoweap Creek. The Canyon's North Rim begins about there and a few North Rim pines could be seen high and far away, up toward the head of the deeply incised creek.

I like Nankoweap. Not just as a name that's fun to pronounce, but as one of the focal points of Grand Canyon history and lore; as a place where man has tried to cope with the canyon.

As might be expected from the name, Nankoweap has an Indian history, though a brief one. In the tenth and eleventh centuries a few generations of Anasazi, the pre-pueblo people, farmed the broad stony-sandy delta at the conjoined mouths of two creeks, perhaps diverting the tepid water of Big Nankoweap Creek to irrigate maize and squash and beans. Their arrowheads and pot-sherds used to be found on the sand up away from the river, but have now been virtually all picked up despite the Antiquities Act and National Park Service regulations. The most prominent reminders of their efforts are several little rectangular rock-slab rooms set into a cliff recess more than eight hundred feet above the river, which they reached, and river people still reach, by scaling a very steep face of shale.

My two favorite Nankoweap stories involve the trail built to the river from the North Rim. Originally it was gouged out in the winter of 1881–82 so that a geological group, of which Major Powell was a member, could study the rock layers exposed at that locality. When they had completed their investigations and departed, the trail was, so it is told, taken over by a concern who called themselves the Grand Canyon Horse and Mare Company. The business of the company was stealing horses, or rather, exchanging horses for a profit, between the Mormon settlers north of the canyon and the Indians who lived beyond it to the south. Stolen stock was driven both ways over the trail, which connected downstream several miles with an old Indian trail to the South Rim, the thirty-five mile combination of the two together soon becoming known as the Horsethief Trail. It should have been a good business because it required almost no investment by the proprietors. With careful mental bookkeeping so that one man's horse wasn't later restolen and driven back for resale to him or one of his neighbors, the "Company" undoubtedly prospered for a time; its business code of ethics was, after all, not too much different from today's.

And then there was the Nankoweap deer drive. Early in the century the North Rim country has been designated a game preserve, game meaning primarily the deer. At that time food chains and ecological balances between plants, animals, and man weren't management tools; game management was carried out with a Winchester. The North Rim deer were seriously depleted and the expedient way of protecting them was to shoot the mountain lions that preyed on them. In 1906, the government hired a bounty warden for the Grand Canyon Game Preserve and in the next fifteen years he virtually eliminated the mountain lion population, several hundred of them. A lion kills and eats at least one deer a week just to stay alive, and with the lion gone, the deer thrived until there were too many. They were competing with range cattle for grazing land and had eaten leaves and twigs and bark of aspens and pines all through the Kaibab Forest as high up as they could reach, standing hind-legged.

In 1924, a local man proposed that he be allowed to

organize a deer drive from the North Rim, where deer were swarming but starving, to the South Rim where they were scarce and browse was plentiful. The drive was to be made down the Nankoweap and up the Tanner, the old Horsethief Trail, by more than a hundred men, who would round up the deer by working their way through the Kaibab pines beating noisemakers to spook the deer ahead of them. They would converge gradually around the head of the Nankoweap, driving several thousand bucks and does ahead of them. Once started down the trail the deer would be contained by it and by the Colorado River canyon until they emerged up on the opposite rim: almost like pouring sand through an hourglass.

The theory and the actuality were as far apart as the mind of man and the brain of the deer. When the drovers finally got together at the trailhead, after being rained on, lost, and generally frazzled by the scraggy country, they found there were no deer ahead of them; all had slipped behind big trees or into rocky arroyos and were scattered back through the Kaibab. Apparently no one had the desire to make a second try; when deer finally did cross the canyon shortly afterward, they were flown over in one of the old slab-sided Ford Tri-motors, just a few fawns who now have descendants who beg pancakes from campers on the South Rim.

Those stories sift back through my mind whenever we come to Nankoweap.

From the slabstone Indian ruins high above the Nankoweap delta, you can look southeast down three miles of beautiful canyon. It is edged at the water with willow whips and tamarisk. It stands on soft insteps of the sloping green gray Bright Angel Shale that has just lifted into view, and it rises as a succession of cliffs and shelves

of Muav and Redwall and Supai and Hermit and Co-
conino and Toroweap and Kaibab, shales and sandstones
and limestones more than three thousand feet in all
above the river and representing more than a hundred
million years of earthtime.

Except for a little jog in the canyon's course one
would be able from the Nankoweap ruins to see Kwagunt Rapid at 55.7, and the Little Colorado River confluence at the end of the long aisle, just below 61. Kwagunt is the antithesis of Badger Creek, which is harder
than it looks; Kwagunt is easier than it looks, provided
the oarsman is always skeptical of this and enters it as
though it were a Badger Creek. The right two-thirds of
Kwagunt is a no man's land of brutal rocks and wild
water and the remainder, beginning left of center, is a
long chain of ragged tailwaves that thrust very close to
the cliff as they are diminishing below, with a well-
defined tongue pointing into them at the head. There is
not as much danger of entering wrongly — because the
holes roar, and throw up spray high enough to be seen
from a boat — as there is of a second's carelessness in
letting the stern swing from its downstream-first position in the tailwaves so that the boat is broached and
rolled, or filled from the side and unmaneuverable near
the cliff beyond the rapid.

The "trail" to the best overlook is through a thicket
of river's edge willow shoots, up through some dilapi-
dated mesquite, and onto a rockpile at the mouth of a
little gulch on the left bank. We reached the rocks hot
and sticky and branch-whipped, but it was a necessary
trip so that the oarsmen could see, and I could see again,
the size of the tailwaves that would motivate us to do
a careful job. Nothing has ever happened, or nearly happened, at Kwagunt, and perhaps as long as we continue

to look it over first, nothing will. The capability of the canyon to build new rapids the waste water can't restructure makes skepticism about each rapid a healthy thing: assume a rapid has changed from last time until you see it hasn't.

It got us wet and threw us' around and scared us about the downstream cliff, but Kwagunt let us have seven good runs and we landed below to bail, and then went on to the Little Colorado for lunch. Normally the "Little C" is a limpid baby-blue because of highly mineralized springs that feed it. One of these, four or five miles up from the mouth, is believed by the Hopi Indians to be their Sipapu, their place of emergence from the underworld to this one. It has been a traditional Mecca to which they have pilgrimaged from their mesa-top villages, a walk of several days from the east. About two miles farther down the Colorado on a wall of Tapeats Sandstone, a heavy precipitation of salt solution is building, as it has for centuries, a crusty white facing of salt over the sandstone, and stalactites of salt wherever it drips from an overhang. A decade ago, Fred, who is very interested in the Hopi and has many Hopi friends, established these deposits as the source of Hopi ceremonial salt, after learning of the landmarks that lie along the salt trail and then hiking and searching them out.

We found the Little Colorado muddy, swollen with upstream rain so that there was no sky-colored water, and looking very much like the caramel-colored Colorado that it hurried to join. It was impossible to cross over to see the little stone "cousin jack" that had been an Indian ruin until Ben Beamer remodeled it under the cliff ledge when he was prospecting and trying to farm there in the 1890's. Across the big river on the slopes of Chuar Butte are still fragments of the two four-engined airliners that fell there after colliding in cloud. That happened in 1956, the first year I ran the river, and an aura of tragedy still pervades the junction of the rivers. There are too many torn pieces of shiny, once-streamlined metal that still catch the noonday sun. The Park Service asks that no one stop there to look around. It's difficult to imagine anyone wanting to.

Two miles beyond the Hopi salt we came to the place where one first feels one is really within the great Grand Canyon that everyone comes to see. The South Rim came into view from the river and on it a tiny dark peg that is really, when one is up there, a tower of brown stone the shape and size of a lighthouse: the Desert View Tower. The sighting of the tower says civilization, people: they are bustling around inside it buying souvenirs and snacks across the counter tops, taking pictures, studying the canyon from stone windowseats; getting out of the sun. Most of them don't know about us, nine miles away and a mile lower, don't even know it's the Colorado River down there. (*"That* little thing? I thought the Colorado was a *big* river."*) But down on the river we know about them, because we've been in the tower and have bought ice cream and gadgets, and have sat and looked down into the depths of the canyon.

We often go back to the rim, to Desert View and the other overlooks, when the trip is ended, to let ourselves understand, again, just where we've been. You must: from the river you can't see the grand immensity that staggers the mind; from the rim you can't find the inner canyon details that the mind can manage. When you have given yourself both perspectives, images of two Grand Canyons begin to melange: the Grand Canyon you look at from the rim because that's all you can do about it from the rim, and the Grand Canyon made of rapids that belabor you, beaches that welcome you, and rocks you can hold in your hand that are a thousand million years old. From the rim the Grand Canyon is scenery. From below the rim it's experience, whether one is trying to grow corn on a dry creek delta, drive stolen stock along a hot dusty trail, or run a hundred major rapids.

It had been a thirty-two mile day from Harding and by camp time we had reached the foot of Nevills Rapid, Mile 75, and had gone nearly as far as we could without entering the granite. In another mile now, slightly more, the reds and gray greens and tans of softer rock would have risen steeply to be high above us, and terraced back out of sight. We would be in a narrow gorge, ragged and black as coal — the Granite Gorge.

The granite is really schist, more than one billion five hundred million years old, a rare revelation of the incomprehensibly ancient sub-stuff of the planet. Once, it was volcanic rock, or sedimentary rock, or some of each, but it has been so violently mangled and thrust that it is no longer either one. The rocks above lie on it horizontally, but the schist stands on end, in slabs and sheets like a continuum of unsorted, carelessly shelved, black pamphlets and papers. No: the *remainder* of the schist stands there; some twelve thousand feet of it above had been eroded away before the overlying layers began forming on it. The schists of the Granite Gorge are the roots of mountains man never saw.

There are really three granite gorges because the schists rise above and then subside below river level from east to west through the park. In all, only about one-fourth of our river miles would be in those sombering roots of old mountains, but it would seem to be much more than that; it always does. You leave the cheering red and tan rocks and slide into the black schist with the same feeling you have turning your back on daylight and starting warily into the dark of an old mine tunnel. The walls of the gorge move in on you; the river hasn't been able to cut a flat-bottomed canyon through the black iron-hard schist, only a ragged vee, and so the rapids become stronger and meaner and the biggest one the river is always the one that's just coming up. You go slowly; even more cautiously.

Part of the dinner conversation had been about the twelve miles of Upper Granite Gorge that were between us and Bright Angel Creek and the rapids in particular, Hance and Sockdologer and Grapevine. "Even if I don't live through tomorrow," said Mary, "this is the greatest thing I've ever done." She was remembering the good times down through Labyrinth, Stillwater, Cataract, and Marble. But she also had the Granite Gorge on her mind, and the bad second day in Marble, and once the morning of the third when the boat she had changed to broached in a trough and she knew it was going over, but it didn't. She and John had, from the beginning, planned to leave at Bright Angel and I wondered how seriously she really doubted she'd arrive there. I promised her she'd make it, and on schedule.

Our run of Unkar Creek Rapid late that afternoon had made me quite sure, as sure as one rapid can, that we were ready for the granite gorges. The entry at Unkar had been touchy, over a come-and-go curler that was hard to find, yet everyone read it perfectly. We had to plunge trustingly into a deep wet trough and know that a lateral current would push out and take us left around boulders; everyone plunged and trusted. It was beautifully done by all, just a few yards out from the right shore with passengers watching and taking pictures, and it made me as proud of us as I had been in Cataract.

If we looked down-canyon early next morning, those of us who had started on the Green — Fred, Maggie, Doug and I — could remind ourselves in terms of days and weeks about how long we'd been on the river. An east-facing crag on the rim of the inner gorge — the gorge that confined us and the Colorado — was lit by the early sun, with a late, full moon just above it. We had watched the sun and the moon share the morning sky in Split Mountain, nearly two months before. A lot of river days, even taken one at a time, the way they must be taken. Somehow the off-river intermissions at Dinosaur and Green River and Flagstaff weren't deducting anything by now; I felt a cumulation of the days from Flaming Gorge as if the expedition had been continuous.

We broke camp and drifted down a mile to the head of Hance.

Hance is the widest rapid. Red Canyon comes tributary there on the left and its boulders have been flood-thrust out into a river that is seventy-five yards wide or more. Some of its boulders had not been carried far when the river current seized them and pushed them downstream and they became a cobbled left-bank promontory that narrows the rapid harshly, farther down. But many other boulders have been driven nearly across the seventy-five yards, or half of it, or a third of it, and bedded themselves, and refused to be dislodged by the river, even the old, strong river. They lie there, some big, some huge, all through it, tearing long shreds of froth down the water, offering ragged boatchutes between, then blocking them with boulders, and the Colorado runs hard and drops a fast thirty feet through it all.

Hance is rated nine or ten on the one-to-ten scale of difficulty.

We always study the rapid hoping there is a way through, yet knowing, really, we'll have to line it because we've had to ever since the dam. There used to be a way through Hance for rowboats at certain levels of runoff, working a completely lightened boat down along the left through the rocks and holes, passing near the promontory that narrows the rapid, and landing just below it. But it was done on natural river, on a volume that doesn't come any more. And a few rocks seem to have changed position. Once again we took the time to look for that old route. Then, not seeing it, or one that would materialize with the lowering water that was on its way, we turned our attention to the lining route.

In lining, the handling of the bow line is very critical. There are usually sections of a lining course down which a boat can be allowed to slide at the current's speed, restrained only by men controlling the strong, hundred-foot line. A boat at the end of its line, pulled by fast water, behaves very much like a kite — a six-hundred-pound one. You can "fly" it in the current and let it down through the rocks where you want it to go only if you're standing in the right place with the other end of the line, just as you can work a flying kite through a grove of trees only by walking the right ground. Fred or Don or I nearly always handle the bow line, with someone one or two steps behind to coil and recoil as it is let out and taken up. The line must always be ready to loop out smoothly as monofilament from a spinning reel; it is zipping out through our hands several yards a second and we are braced against the boat in the current. If a coil snarls behind our hands, we can be jerked out into the river — or let the river have the boat; a choice that is not at all clear-cut, for there are times when one must make surprising choices to maximize or minimize, or make possible or make probable. A dependable coiler is insurance against having to make such choices.

Don and Fred and I became the cadre for the first let-down, with Doug, Dan, Chuck, and Walt wading into the water or crawlng out over the rocks, to help the *Norm* past the difficult places. Don and I took the bow line, he coiling, and Fred boarded the stern deck to be with the boat when it grounded. We soaked the coil so the line wouldn't burn our hands going out and Doug and Chuck, life-jacketed as were we all, walked the boat out into chest-deep river bottomed with rounded, slippery boulders. "Ready?" called Doug.

"Ready out here," answered Fred.

"Ready, Don?" I asked over my shoulder.

"Ready."

They pushed the *Norm* out a few feet more and the current began to purchase it, gradually at first, then more strongly, drawing it down toward a cluster of rocks. We were on hand signals now because of the roar of a hundred holes out in Hance and the distance between us. I squeezed the line a little more, increasing its drag through my fingers, feeling it instantly warmer in spite of its wetness; the boat slowed a little; Fred's arms went straight up; we braced and held. The boat stopped just above a mid-chute rock. He dropped his arms, then circled an index finger over his head slowly and we eased him down until the boat stopped against a rock. Fred got out on it, pushed the boat past, and got back on the stern. We let him down a few more yards; sitting with his legs over the transom he kicked away from one rock, then another, then another. Within a few dozen yards we had let the *Norm* down into a pool that was fed by

Lining the heavy boats is hard work, done in fast shallows, on slippery footing. The objective is to substitute scraped shins and scuffed chines for more serious consequences such as the loss of a boatman or a boat.

a fast chute, and drained by another that left it laterally toward the rapid. We grounded the boat there on submerged rocks with Fred holding it in place; then Don and I moved down the shallows along shore, he re-coiling the line as we went. We worked our way out above the fast pool onto rocks that gave us our angle of control. Ready again, checking everyone, we held the boat, its pointed bow parting the thrusting current as Fred set it away from the rock, not going with it this time. I played the boat in the current until it seemed properly positioned, then let it run and, as the stern flanked the lateral chute, whipped the bow line; the boat swung and the side current caught it and took it into the chute, and down around the corner, and grounded it on a large flat rock, barely covered. Fred and several others tugged and skidded it over into the head of a pool below, standing in swift water on dangerously slippery rocks. While that was being done, Don and I repositioned and re-coiled the line, taking a station on big rocks overlooking the boat, as close as we could for maximum letdown length in the next step of the operation. Now the boat would have to run the length of a long, fast pool, be "flown" directly into the head of a crooked chute going back out into deep river, then when it washed through that, be pulled into the lee of rocks just below. Having gotten it off the large flat rock and into the head of the pool, we let the current have it, but it was pushed down and against the shore side of the pool several times before we found the right water, and each time had to be dragged back up against the strong current and swung out again. Finally it caught the proper zone of the fast water and, as it came opposite the chute, a whip of the line spun the stern to the head of the chute and the leaving water caught it and swept it through into calmness below, with only a few feet of bow line remaining in our coil. The boat was now just above a bouldered promontory, in the lee of eight-foot rocks. There was no current to play it in, and we could only hold it there until one of the men climbed out through fast chutes and rocks to bring it in against the left bank. There the oars were replaced and a run of a few dozen yards made along the shore as closely as possible, down to the lower "beach," which was actually a coarsely bouldered shore. The river, reacting to its confinement by the promontory, surged badly there and it was necessary to haul the *Norm* and each succeeding boat onto the rocks above the lifting-dashing action of the waves for the reloading.

Lining the *Norm* took forty-five minutes. We made one or two small changes in the lining course and the next boat took forty. As we became practiced, and the others worked into the team, we reduced the time to twenty-five minutes per boat. Meanwhile our passengers had carried not only their own gear but that of their boatman as well down to the lower beach, over difficult boulders and a high, hot, sandy bench. We finished the last boat just as receding water was beginning to make the smoothly working procedure more difficult, about noon, and made lunch before starting into the Upper Granite Gorge.

The strata come up sharply in the vicinity of Hance, and the gorge begins where the tailwaves are still dissipating. The first mile contains some rather lyric geology and I thought I might work out something to ease Mary's mind, but all I could contrive was:

> *Shinumo Quartzite,*
> *Hakatai Shale;*

A mile to the Granite,
Get ready to bail.

and I decided to keep it to myself. In a mile the Vishnu had pitched up to become the somber foundation of the other layers; hard, shiny black, imposing crooked narrowness on the river and the boats. Little grows on it. There is no soil, no flatness. No room even for scraps of shore. Black cliffs right to the water. The Granite Gorge felt again, and always will, like an ancient place, a Genesis place.

Twenty minutes after leaving Hance, we made the sharp turn of gorge to the right that is the advance landmark for Sockdologer, and then could hear its voice. "Sock" has one of the louder voices in the Grand Canyon; it is one of the larger and longer rapids. Powell named it: sockdologer, a corruption of the word "doxology," was a popular slang term of his day for something unusually large or decisive. But even if one's vocabulary didn't advise a stop to reconnoiter, one's ears would.

Staying well to the left side of the river to avoid the thrust of the main current, we nosed the boats into quiet water in the lee of projecting slabs, and tied. The shoulder of the gorge was steep and the dark, indurated rock full of midday sun and too hot to press against; we could even feel it through our shoes. Hand-and-footing our way up to a vantage point was impossible; we had to keep our feet under us and scale the wall with no more than a quick push at the hot rock when a foot slipped or we lost balance. Dislodged slabs and hunks hitting the bedrock below clanged as if they were falling on iron. When we had gone fifty feet up, and a few rods downstream, we could see Sock spread out below us.

"Hasn't changed any," said Fred. "Still big as hell."
It was. There was the big tongue, pouring down in mid-river into a backbone of immense wavestacks that ran for three hundred yards before washing out in the quieter water; left of the stacks, river badly torn and thrashed by ledges and boulders; on the right, the wall pushing in its great breakerlike rolls of water much higher than a boatlength from trough to crest. Right of center, between the backbone of waves and the breaker-like rolls, we could point out an aisle of lesser waves. That aisle is where we usually run Sock, and it seems running through there as if the whole rapid is above you.

The water was moving off the tongue to the right so that it swerved into the right-side rollers. To avoid them we'd have to stay left on the tongue, a tactic that was complicated by a deep lateral hole coming out from the upper left; a hole we could not see into but could infer from the plunge of water over a sharp brink and a trail of froth angling out just below. We studied the course a long time so everyone would feel it was critical, watching occasional pieces of wood from a flashflood somewhere float down the tongue at different places, saying "a little farther right than that one," or "about a boat-width left of where that one went," until everyone had his own visual measurement made. Then there came a moment that could be felt, when the talk all but stopped and the fidgeting increased, a time when everyone was ready and if we waited much longer they wouldn't be. We eased ourselves back down over the burning, clanging rock hunks, stopping partway down to see the head of the rapid more nearly as it would look from the boats. When we had returned to them the hole and the froth were completely out of sight below the brink.

For the first hundred feet we stayed left of mid-river,

unable to see anything beyond the smooth horizon of water where Sock started its long rampaging twenty-five foot drop. The roar was awesome and the tawny water pulled us toward it between the black walls. The last tailwaves started coming into view, and then in a few seconds most of the rapid, slanting down and away, a field of wild water not showing much of the pattern of stacks and rollers that had been so obvious from high on the rocks. The mound of water just above the hole then appeared, and beyond it the line of foam trailing obliquely out onto the lower tongue. The separate sound of the hole surged into focus and we moved slightly left to take just a little more of the mound above it — then we were over the mound and through the foam and down the tongue and into the waves and had found the long safe aisle and everything in the rapid seemed to be rising above us as high as the gorge walls themselves and about to crash down and fill us but none of it did, and we bobbed out below, wet, but with only three or four inches of water aboard. Looking back, I could only keep track of the other boats by their different distances from the *Norm,* and none was in sight except when it rose on a crest. The lower part of Sock has historically been a troublemaker, and only when all boats were beyond the tailwaves did it seem safe to relax.

The hot sun nearly dried us before we had drifted two more miles to Grapevine Rapid. "Grape" drops only sixteen feet, but is wetter and longer than Sock. To study it we again had to climb the blistering black schist, from where we could look down on a tongue that was truncated by a row of holes just above the beginning of the tailwaves, and right and left sides that were torn into turbulence by jagged, close-set walls. The foot of the rapid ran partly into, partly past, a black buttress of the

*In the black schist of the Granite Gorge. The crest of
each big tailwave puts a boat momentarily in sight . . .*

*. . . then it is gone, hidden in the deep trough that
always lies below.*

left wall, and beyond that the tailwaves ran on, out of sight around a corner. Grape is easily a half mile long.

Standing above Grapevine we began reviewing the run of Sock, our first chance to do so. Some of the men said they had recognized a number of the waves as they passed them, which was encouraging; some of them said they couldn't tell very well where they were in the rapid. But our Sock run had been one of our best. Looking back for them as they came through, I had seen that each was where he should have been, and it was dramatic to watch them come, each in turn, through the last of the tailwaves. If we could get seven-for-seven through Grape we should have no more problem rapids that day; Bright Angel Creek was just six miles farther.

We decided to slip off the tip of the tongue to the right, taking on a wet lateral wave to miss the truncating holes, then to quarter slightly left and pull for the middle. There was a zone of relatively smooth water between the tongue and beginning of the tailwaves in which we calculated we could get three strong quartering strokes before having to turn the stern back into the turbulence. After that we could only ride it out, a wave at a time.

I didn't see them again, except the *Camscott* right behind me, until we were a mile farther down the river. The buttressing schists cut off the view and the tailwaves seemed to go on forever, but when they had finally subsided we found an eddy and waited. One by one, six boats came around the corner, bailing as we had done. When Doug came into sight in the "tail" position we moved out and caught the current again.

An hour later, after running a half-dozen rather wet riffles and Zoroaster Canyon Rapid, we rowed into the big eddy just above Bright Angel Creek and tied for the day. It was three o'clock. Rummaging through our duffel we found our wallets and a few other necessities and trudged to the back of the beach to intersect the Kaibab Trail.

Bright Angel is a capricious little creek that rattles down from Grand Canyon's North Rim on the cobblestone floor of a narrow side canyon. About a mile from the river, the creek swings hard against the west wall of the canyon and in this temporary wideness is Fred Harvey's Phantom Ranch. It consists of a few cabins and a dining room laid up of Bright Angel boulders and dark-stained wood, with eaves and sashes of green; a swimming pool fed by the creek, an adjoining bathhouse, some shower rooms, and down the canyon a short distance, a barn and corral. Anywhere but in a national park, Phantom would have to be boarded up, or bulldozed and replaced by a five-hundred-room lodge of glass and prestressed concrete. It is wonderfully out of touch with the times: too old, too plain, too small, too expensive, too far off the beaten path, too hard to supply. It is attractive, and an enjoyable place to be because it doesn't try to overwhelm the canyon; it merely nestles in it, in a grove of big trees. A meal at Phantom costs nearly as much as a cabin overnight because every pork chop, every can of corn, has to be packed down on the back of a mule over eight miles of steep, dusty trail, and then later the empty can packed back out to the rim for disposal. But the meals are some of the best dining-room meals there are.

You can't drive to Phantom. You hike or ride a mule. The trails from the rim are three or four feet wide, edged with chunks of whatever rock strata they happen to be traversing at a given elevation, and concaved by the scruffing of tens of thousands of hiking boots and mule hooves. The easiest walking is in the middle where the trails are worn deepest. The steepest of the two, the Kaibab, starts at Yaki Point and virtually plummets down as almost continuous switchbacks, but it is shorter and faster. The Bright Angel Trail, which starts in the South Rim village, is five miles longer, and therefore more gradual. At the mouth of Bright Angel Creek the trails converge, then follow the creek up to Phantom.

Phantom's pear-shaped swimming pool is its focal point, its coolest, shadiest place, and it collects people from the trails, the cabins, and the river. "Phantom people" are a select assortment; they have experienced the canyon in one way or another. The faces at Phantom change from trip to trip, but the types don't.

There are the wranglers, waiting to start "to the top" with a mule trail in the morning, or resting from just having brought one down. A number of them are in their twenties, lanky Mormon boys from southern Utah and northern Arizona. They wear sideburns and slightly stagy western shirts, Levis with sharp-toed saddle boots, and they bend their straw-hat brims hard up against the crown on the sides and low to a scooped point over their eyes. Perhaps on the trail the boys slide their hats to the backs of their heads to let the sun under them, for they don't have the white forehead line that the older wranglers have. Old and young, the wranglers are all good-humored men, as anyone who works with mules must have to be. When they change from their long-sleeved trail gear to swim, they are absolutely white except for deeply tanned faces, necks, and the backs of hands. A lot of their poolside banter is directed at the day's happenings on the trail and perhaps the tribulations of their riders in getting comfortable on a saddle mule a yard wide.

There are hikers, some alone, some in twos and threes, propped against the bathhouse wall with their packs as backrests. Some will have hiking boots and heavy socks pulled off (at last!) to let their feet swell or to check for blisters. They wear blue jeans or hiking shorts, tee shirts, battered old cloth hats from surplus stores, a sizable canteen or two, and perhaps a bandanna tied across their forehead to keep the sweat out of their eyes. The proportion of foreign visitors among hikers is high; they are more accustomed than Americans to walking, and having come so far are not content merely to look into the canyon.

By midafternoon, riders have usually arrived on mules to spend the night in one of the little cabins and ride out after breakfast next morning. Their feet are all right but they are wondering about the area that was exposed to the saddle, and are ready for a swim.

And just about any day between Easter and Halloween, there will be river people around the pool. River travel through Grand Canyon has increased almost unbelievably; ten years ago fewer than five hundred had made the trip, now twenty thousand have.

John and Mary Womer, Rosemary Philips, and Tommy Dunlavey were leaving the expedition at Phantom, to be replaced by Mike Wiedman, Marty Richmond, Don and Cherie Dedera, and Doug's fiancée, Sue Cunneen. As we had left Lees Ferry with one empty seat, this gave us a full complement for the last section. Mike and Marty had already been at Phantom a few days, hiking some of the inner canyon trails. Sue arrived late in the afternoon, having wisely waited until the hottest part of the day had passed before starting, and then having come down the Kaibab Trail to save time and miles. A call to Joan on the South Rim determined that duffel and new provisions had gone to the Yaki Point trailhead for loading on pack mules about daylight the next morning, and that saddle mules would be available for the four from the river to the rim.

The rest of the day was very pleasant, the Granite Gorge was a million miles away and everything was arranged for the last ten days. I had a cold beer with Jay Goza, the boss of the wranglers, while we puzzled over the wiring of one of the pool's filter motors, read a few pages of a paperback I'd been carrying in my camera case, and wrote a half-dozen short letters to people I thought would enjoy receiving something postmarked, "Mailed at the bottom of Grand Canyon." When I'd squeezed everything from the evening that I could, I found my way back down the trail in the dark and went to sleep beside the *Norm*'s bow line.

The pack mules arrived at seven with perishables for the remainder of the trip boxed and wrapped in canvas, lashed to the packsaddles. We unbound it all and checked off twenty loaves of bread, ten dozen eggs, a dozen heads of lettuce and a dozen of cabbage, a lug of tomatoes, twenty pounds of bacon, and dry-iced packs of chicken, steak, and pork chops. The canned goods and staples were already aboard, having been brought for the entire trip from Lees Ferry. By eight the sun was already bearing hard on Bright Angel Beach and the hot inner canyon air was motionless. Loading the boats became hot, sticky work made more difficult by the angle at which overnight low water had left them on the steep face of the eddy-cut sand. Gradually more water began arriving — the previous afternoon's computerized increase at the dam — and by midmorning they were floating level again and we had them loaded. At the ranger's suggestion we carried empty canteens back up the trail to the Park Service cabin and filled them from his outside hose tap, then distributed them among the boats. The new passengers filled in the empty seats and we snugged up life-jacket straps and pushed out into the eddy.

Now it was Phantom and its pool that were a million miles away, and once again the river and the granite became everything. Twenty-eight miles of granite gorge lay between us and the roseate rocks that again reached river level near Elves' Chasm. In those twenty-eight miles are:

MILE	RAPID	DROP
90.3	Horn Creek	8 feet
93.4	Granite	17 feet
94.8	Hermit	14 feet
96.5	Boucher	13 feet
98.2	Crystal	17 feet
99.2	Tuna	10 feet
100.5	Agate	5 feet
101.1	Sapphire	7 feet
102.6	Turquoise	4 feet
104.7	Ruby	8 feet
105.9	Serpentine	10 feet
107.6	Bass	4 feet
108.5	Shinumo	8 feet
109.5	110 Mile	21 feet
110.3	Copper Canyon	5 feet
110.9	Hakatai	8 feet
112.1	Waltenberg	14 feet
114.1	Garnet	2 feet

Of Grand Canyon's twenty rapids most problematic to rowboats, five are in that relatively short distance: Horn, Granite, Hermit, Crystal, and Waltenberg. I was especially anxious to get beyond the first two so that I could even a very personal score with Hermit; it had cost me a day of lost time the previous trip, time that we had planned to use to hike to Thunder River, and I was still smarting from the idea of the setback. As we had come down the river, I had found my spare-moment thoughts going more and more to Hermit and now with seven miles to go, in my mind I had already run Horn and lined Granite and was keyed to Hermit. I thought about it down through the Devil's Spittoon and the rest of the fast water that began at the mouth of Bright Angel Creek and ran for two miles almost to the head of Horn. Then Hermit had to be pushed back again and Horn, for the time, had to become the only rapid on the river.

Horn Creek is almost too risky for passengers to ride and we started them along the river shore, offering them as partial compensation a chance for pictures of Horn, one of the most photogenic rapids in Grand Canyon. Halfway down, a black jut of bedrock thrusts out under water to create an indescribably vicious lateral hole that swallows most of the river, and then regurgitates it boiling and fizzing into a few tailwaves. One can photograph from the black rock just out of reach of the spume exploding from the hole, with the boats dashing past thirty or forty feet away through a foreground of wild water and a background of somber gorge.

While our passengers were making their way down, we climbed higher on the wall to satisfy ourselves about the entry over a mound of water into the left side of the rapid. Then the *Norm*, followed by two other boats, set out to make the proof run while the other four oarsmen studied it for workability. We found it to be as we had read it from the cliff, an exciting run, very close to everything dangerous, but with margin enough.

By riding back up the eddy current afterward we could float in the lee of the jutting bedrock and watch Fred lead the second column of boats through. We could see nothing of them until each came, in turn, to the place of the Moment of Truth, the last inch of level water, higher enough that we looked upward to it, to the boat graceful and crisply white against black walls and brown water. It poised there a second in profile; then it began sliding down a sleek chute so steep you could see into its green-painted cockpit. For a moment it disappeared behind a wall of water into a deep trough above a savage hole, and when it rose out, it was being carried around the hole by a variant of current we had counted on, and seemingly committed to the lateral hole that swallowed most of the river. Then, on the shoulder of a prechosen wave, each oarsman dipped deep and pulled hard, and his boat moved slightly left and plowed only the extreme left end of the hole. The heavy wave smashed against his stern, swamped over the deck, and was thrown aside by the vee-shaped splashboard, as the boat joined the others in the eddy.

We lined Granite Falls that afternoon, as most of us had expected to do, and made camp opposite its tail-waves. Now, finally, Hermit was next, and I could think of it as the only rapid on the river. I had been waiting for another chance at Hermit for nearly a year.

On our last trip, the summer before, we had arrived at Hermit late in the morning. The river is lowest at Hermit about dawn, increasing constantly to a midafternoon high, then decreasing again to the next dawn low. We were too late that time; the rise was well under way and the rapid already too formidable. Experience told me it couldn't be run at that stage of water, but several of the passenagers — and one or two oarsmen — weren't convinced. It looked easy; Hermit is a straight run, without rock hazards, down a fine, unbroken tongue and through tailwaves into quiet water. I was remembering how much more immense and powerful those tail-waves are than they appeared to be from shore. But my "Those waves are much bigger than they look from here" didn't satisfy the thrill seekers among us.

Then, propitiously, a passenger-crammed pontoon boat had appeared around the bend of the river and, without stopping to reconnoiter it, had driven into Hermit. Wallowing through the great waves, the thirty-three-foot rig gave them scale: it showed the wave length from trough to crest to be nearly forty feet, for one end of the pontoon was being pummeled by the exploding crest while the other was still deep in the trough behind. The big rig plowed into the wave, shuddered — hesitated — spume poured in over the bow; it flexed and heaved; arms and legs and hatted heads could be seen jumbled in the spume like jackstraws. Then its yowling outboard motor shoved the pontoon through into the lesser waves below, as everyone aboard started untangling themselves from each other. The proposition of our running Hermit in a boat only half as long and a third as wide was put to rest more eloquently than I could have ever put it with words.

We had continued to wait, that time, for the afternoon decrease to bring us a runnable stage. The day had dragged on slowly, with little for the others to do but read or nap or explore the immediate vicinity while I watched the river, particularly the first tailwave, the forty-footer. When it had lost enough of its crashing crest we could go.

As the day tapered into late afternoon I wondered whether this would happen, but by the dinner hour it seemed to be starting and I passed the word to hurry with dinner and dishes. We ate in the heavy inner canyon dusk with sunlight fading on the rim a mile above, and I walked down once more and looked, and thought the force of the wave crest looked marginal. It would clean up more in time; we would have to give it all we could, and go in the last usable minutes of daylight.

Saving just enough to run the rapid and perhaps make another mile before darkness, we had pushed out, nervous, nearly talkless, making only the terse scraps of sentences essential to boarding and checking everything, and had entered Hermit in closer than usual column.

The great tailwaves were all that anyone remembered of that run, grand, glassy mountains of water with swooping valleys between; climbing for three boatlengths or more until it seemed we'd lost all power to climb; perching then on a crest as sharp as an alp and wondering whether we'd pitch forward into the next trough or back down the slope onto another boat (everyone saying later they looked up and saw the whole bottom of the next boat out of the water above them on the crest ahead); pitching down into another trough nearly as deep and soaring up to another crest nearly as high, and after four or five crests finally being sure we would carry over the next, not fall back. And looking around at each other when we reached the still water, almost amazed to see all boats right side up: it was supposed to work that way, but had still seemed incredible. After that we ran Boucher, an easy rapid, able only to hear it all around us and to detect only glassy highlighted hints of wave shapes against the blackness in the gorge, and landed two miles farther down in darkness that was absolute, on a small beach just above the roar of Crystal.

So we had run Hermit that other time, but it had cost the day. The following winter, going over the Bureau of Reclamation flow charts, and the measured flows and observation schedules of the Geological Survey at Bright Angel, I thought I'd found a way to cope with Hermit: since we had been able to find a runnable stage at the end of the day with water falling, I reasoned we should find a comparable stage at the beginning of the day with water rising. And using the morning rise should give us time to choose an even more suitable stage, without our being pushed against darkness. It was to try this approach to Hermit that I had been waiting for so many months.

Now it was morning; at least it was light enough to see that the river was down and the boats had been left high on the sand by the overnight low. I had spent the night, it seemed, skirting the edge of sleep, waiting for that first light that would let me check the flow level of the river and let us get started down to Hermit. I was awake when the sky lightened almost imperceptibly up the canyon to the east, and shook the sand from my bed and rolled it and stowed it in the *Norm*, with enough noise to awaken Fred, who was always first up, every morning but this. Then quickly he was there and we banged and rattled around the kitchen area to get the others up and moving and when coffee had just rolled, sent false-alarm shouts of "last call for breakfast" ricocheting against the canyon walls. Finally the stragglers shuffled, puffy eyed, to the fire. They seemed in no special hurry and there was no way to really make them understand the urgency of my appointment with Hermit. By now it had become a highly personal thing.

By seven we had rushed everyone through breakfast and moved down a mile and a half to the head of the rapid, to the pooled water above the roaring voice. We went down to look. We walked, but inside myself I was running all the way, first to a massive boulder overlooking

Portrait: a Hermit wave

the rock-narrowed tongue, a formality, then a hundred yards farther to the all-important tailwaves. They were lower than they had been just at dark a year before, but steep and mean, and the river rushed at the base of the first with such force that it was nearly knocked over, the crest toppling back upstream thrashing and falling down on the sleek water climbing to it. We were early enough, I decided, and sat down on a close boulder to watch the wave change.

The sun was getting into the canyon now and you could smell the heat of the Granite Gorge beginning to build for the day, the flat heavy heat of early August. After a time, when I looked away from the wave after studying it intently, I found the motion of the river had dizzied my eyes and the black schist on the opposite bank appeared to moil and flow within itself, eliciting a vertigo that made constant staring at the river more comfortable than looking away. When an hour had passed, higher water had begun to push back the toppling crest, trying to smooth it into a runnable wave, and those who had stayed at the boats thinking we would take a quick look and go had begun to wander down in twos and threes to see how much longer it would be. I walked up to the big boulder near the head to recheck the entry and course above the wave, making a boat of my eyes, letting the river carry them through, letting it pull them where it would pull a boat, entry after entry. But something was missing; they always ended their run in the deep hole to the left of the wave. Dan was there and threw in driftwood for me to watch, and it, too, was carried inexorably to the hole and swallowed. I climbed a buttress of schist back against the cliff for an overview, looking for the narrow nuance of current that, other years, has always taken me a few feet to the right, past the hole and over the less ragged shoulder of the wave, but it wasn't there, for my eyes, or for the sticks he continued to sidearm onto the tongue at all the right places.

I went back down to the rapid to look for it and felt the sharp stab of uncertainty about my plan. The river was swelling now; computerized water was coming stronger and had cast the first wave into a smoother, runnable mountain. But the river force that wave had been absorbing was sweeping through it and plowing into the base of the second wave, making of it what the first had been. Below was the third wave, and the fourth, and a half dozen more. Would this dominolike effect now have to work its way down through all of them? Had Hermit done this on the evening end of the cycle a year before? I didn't remember it. Had it been so dark; had I been so keyed up to run that I just hadn't seen it? Why wasn't my little right-swerving current in the rapid this time?

And now it was late morning and everyone was getting restive.

Alone, I might have run it, or tried to, in time, trusting the saving little current that just had to be in there, somewhere, that I was still searching for like one of the drawings hidden in drawings in the children's game pages of Sunday papers. Perhaps, once in the rapid, I would find it, or it would find me. My oarsmen say I make running a rapid look too easy. But many times I feel my boat is an instrument, and a difficult rapid my opus, with every oar stroke a note being played in time and on key, and I'm making music through the wild water. That delusion, if it is delusion, comes to me only during good runs of bad rapids, when time is suspended and there are only distances and dangers, and my arms have become oars that mostly move themselves when they should, and put the boat where it should go.

But there are other times, too, and always will be, for the river is always in charge, and one never can say that one has mastered it finally. During those times — I think about them at the head of rapids like Hermit — I've played wrong notes that turned my sweet, private song into sour, frantic tries to refind the intended melody, and once or twice, failing to refind it, I've been capsized. Some of the passengers and some of the boatmen, impatient to get on down the river, were willing for us to take our chances at running Hermit, right now. It's happened there before, and at other rapids. "I'm a good swimmer," they'll say; "let me try it." And once: "Look — I'll buy one of the boats from you, here and now; if I don't make it, you're off the hook." You can't defeat this thrill-seeker syndrome with logic; you can only give a flat and final No and endure the hard feelings. This is the leader's greatest anguish, mine at least: if I pronounce a rapid unrunnable and allow the matter to go into discussion, then it's up to me to prove unrunnability by trying it and getting into trouble. If I don't imperil myself to prove it so, then there usually are those who will always think it *might* have been possible. But one gets these built-in judgments, subconscious measurings of a rapid, and one learns to trust them. One boat might make it. Or two, or three. But in a marginal situation the probability of seven successful runs is not high enough.

So it had to be called unrunnable this time. By lining, we might still salvage a half day, and having to settle for that, for now, I gave up my lonesome proposition on the boulder beside the wave and poked my way reluctantly back upstream to the group.

We worked out our moves at the critical points and began, down the left, into a fast, crooked chute, around a sharp hook at the bottom that dropped over a ten-foot choke rock into a boulder-locked pool. First the *Norm,* rowed to the head of the chute, held in check by two oarsmen walking the bow line along shore. Then held while oars were shipped and the cockpit secured. Let into the rapacious current of the chute, checked by a momentary squeeze of the sizzling line to let a dogleg of rushing water take the stern in past an outside rock, into the lee of another rock where the boat could be caught and held. Pushed out again with the bow line recoiled, seized again by the current of the chute, shooting on and around a severe hook and over the water-sheathed choke rock to plummet with a splash into the quiet water of the boulder-protected pool beneath, and tied there.

And then the *Camscott.* And then the *Sandra,* and then the *Bright Angel,* and suddenly, trouble! They didn't slow her momentum enough with the bow line to give her time to make the dogleg; her stern crunched against the rock and stopped her and her bow went over against another rock and she impinged. With all the force of the chute jamming her against the rocks and filling her, she went down in the chute, two feet deep, just deeper than she, and she bottomed and lay there, the water etching a ragged boat shape in the chute as it rushed across her decks and filled the cockpit.

"And there goes the day," I said to myself silently, feeling again, for a second or two, the loneliness that comes right at the brink of a bad one, or wave-watching at the foot of another; the times when no one can help. I knew I should have handled the bow line for the let-downs into the chute.

She was like a wooden shoal in the fast current; stepping on her decks in the few inches of fast water sweeping across them, we could make our way to a cluster of huge rocks farther out. Locating and hanging to her safety lines, we managed to tie other lines at stem and stern and passengers and oarsmen teamed up on them, heaving from several experimental angles, from shore side and then from the other rocks out beyond her, without budging her. Someone found a plank up Hermit Creek and we tried jamming it against the bow and into the rocks of the chute bottom as a lever. On the count of "one, two, three — HEAVE!" the lines went bowstring taut; the planks crackled warningly, but she still wouldn't move.

She wasn't lost, yet; lowering water that night would likely drop the river below her gunwale so she could be bailed and lightened enough to refloat. But she was blocking our only lining chute, and there were still three more boats to bring through. Now it was midafternoon.

And now a pontoon party landed at the head of Hermit and the boatmen, coming down to reconnoiter the rapid, found us struggling with lines and poles, prying and heaving at the underwater shape, and went quickly back to get the others, and their cameras. Within minutes, fifteen or twenty strangers had clustered around, peering at the water-swept outline of the boat, asking their boatmen, "who are they?"; "what are they doing?"; "does this happen all the time?" Then, after a while, they went back to their pontoon and ran Hermit, yahooing and waving in an overdone show of enjoyment for our benefit as they wallowed through the hole and the ragged waves and went on down the river.

Not knowing what else to do, I went out to her, through the pushy water, hanging to the taut stern line, and, reaching underwater to the portage bar across her transom, tugged tokenishly. And she moved — began sliding, scruffing along the bottom, cleared the dogleg, bumped down for a boatlength and stopped again, still submerged, in easier water.

There is no explaining why that half-inch, half-hearted tug had dislodged her when lines and levers wouldn't, but now she was in the lee of the dogleg and lying with the current. Chuck lowered himself into the chute, got his brawny shoulder under her bow, and like an Atlas, shouldered it high enough so that the river no longer swept over her, but past the gunwales, and somehow held her until enough water could be bucketed from the cockpit so that she rose to clear the surface full length.

Now she was buoyant. The current wanted to seize her again, but she was held by a score of determined boatmen and passengers, her bow and stern lines snubbed around nearby boulders. We had her back now, and we bailed and sponged every bit of water from her before we let her go with the thrust of the chute, down around the sharp corner, over the big choke rock, to splash into the pool below. The last three boats were lined with more caution at the dogleg section of the chute and lowered anxiously into the pool, then reloaded with supplies and duffel. We moved on downriver and went into camp not far above Crystal Rapid, not long before dark.

Around the fire, after canned cocktails and a good dinner, we began getting Day Thirty-six back into the proper perspective: it had been an experience, not an ordeal. To Powell's men a hundred years before, such a day would have been routine, and they would have ended it with nothing more than a few bites of dried apple, moldy-flour biscuits they called "gods" (probably an abridgment of a profanation), and a cup or two of strong coffee. We had soup and salad, pork chops and green peas, canned peaches and tea.

A hundred and thirty miles to go now, with about sixty rapids yet to run, most of them threes, fours, and fives; just good fun. But there were still a half-dozen tough ones, as dangerous to line as to run. The first of these, Crystal, now lay just ahead.

Crystal is a new rapid. Until the winter of 1966–67 it was only a long, easily run riffle, rating two or three on the ten-point scale. Then an extraordinary rainstorm fell on the Grand Canyon's North Rim, sending a monstrous flashflood rampaging down Crystal Creek to the river, a head of storm water so high and powerful it ripped out thousands of tons of boulders — an entire boulder bar thirty feet thick and a hundred yards wide at the mouth of the canyon — and swept it into the right side of the Colorado, forcing the river over into a narrow, hole-studded chute along the left. The river, because Glen Canyon Dam captures its strong spring runoff, now lacks the power to rearrange these rocks, and Crystal is the worst rapid on the river.

One can't help feeling that the lining of a rapid counts against one in the running of a river. It's a kind of demerit; it gives him less than a perfect score. Experience had told us we would almost certainly have to line Granite Falls, Crystal, Upset, and Lava Falls. It had been a disappointment to unexpectedly have to line Tanner, back at Mile 24.5, a setback in my self-imposed tally of rapids run and rapids lined. Then Hermit had added another. Now we needed to run two rapids that we usually lined in order to even things again, and I looked at Crystal ambitiously.

Pontoon boats run Crystal; they have to, and they go through bucking and heaving, diving into its spume-

Jack Currey drives a pontoon boat through one of the holes in Crystal.

filled holes and slithering over its exposed rocks. But different things are possible in a pontoon boat than in a Cataract Boat, just as different things are possible in a jet than in a sailplane, and I could see that if we were to make up for Tanner or Hermit, it wouldn't be at Crystal. Not this time, anyway. At some stages of water, a Cataract Boat *might* sneak down the narrow chute, as close to the crowding right-bank rocks as possible without touching one and being thrown out of position, *might* avoid the first monstrous mid-channel hole, then *might* pick its way down through rocky right-hand shallows where the river regains its width below the great choke field of boulders. But it would have to be done without passengers, and any boat that made a bad run would either miss its landing and leave its passengers stranded at the mouth of Crystal Creek, or would be swept into a field of vicious holes and shoals that would tear it and its oarsman apart.

"Maybe Deubendorff, then," I said to myself, thinking ahead thirty miles to the next running/lining decision. The oarsmen and I made our way to the head of Crystal, working out our lining course and procedure as we went.

We had two boats lined through, and a third one down to a point just above the first big hole, when it slid into a too narrow chute and caught, the stern pinched between rocks at either side, a routine occurrence in lining. I made my way out to her through the shallows, braced my feet in underwater boulders, my chest against the boat and lunged against her. Something popped dully and deeply in my abdomen and a sharp pain shot through my chest, and I knew I'd cracked or broken a rib. I felt momentarily weak, and breathing hurt. Not wanting anyone to know, I leaned over the boat for half a minute or so, reaching down as if exploring the contours of a rock that was holding the boat. The pain continued, dull between breaths, sharp when I inhaled. "I guess I'll need a little help to get her off," I called to the oarsmen on shore, and Chuck and Don came out and lifted her free, I only pretending to help. "If you two and Dan can take her on down, the rest of us will go get the next one started," I told them. But instead I avoided the others and went back up to the head of the rapid and found a place out of sight in the scanty shade of a tamarisk bush, and lay down for a few minutes. Suddenly, I had had enough. Thirty-seven days, five hundred and fifty miles of river. Was it the accumulation of days, or miles, or the incessant sun in the black gorge, or the sand always in shoes and bedroll, or my chapped backside from the always wet boat seat, or the resented linings, or the undisguised triumph of the pontoon people as they ran the "unrunnables," waving at our lining crews, or the pressure I put on myself to do better than the others so I could expect them to follow me down the river? Or was it just the pain of a cracked rib? Everything was intertangled, and it was impossible to tell cause from effect. And seven more days. When all the boats had been lined through Crystal, would I be able to row strongly enough?

While I was resting and wondering, two of Jack Currey's pontoons came down the river and landed nearby, with a party of thirty or forty people. Jack and I had last met at Hell's Half Mile on the Green, two months before. "Got something for you," he said. "Joan sent it down with me." He went back to one of the boats and brought back several quarts of ice cream, still nicely frozen after five days on the river. He was obviously but quietly pleased with himself that his big pontoon boats could carry frozen goods down the river to our hot little group, and I couldn't begrudge him the feeling. As we stood talking, I held the armload of frosty cartons against my rib, their coldness dulling the pain somewhat.

When they had gone, I carried the boxes of ice cream down to the little beach where the already lined boats were tied, and we interrupted the lining for an all ice-cream lunch, everyone eagerly receiving several thick slabs of vanilla in his big metal cup, surprised and pleased at the unexpected treat.

ONLY THE RECKLESS FEW

I had knotted two dish towels together and bound my ribs, wearing a shirt again to conceal the binding. Rowing hurt, but at least it was possible, and after Crystal there was easier river. We emerged from the Granite Gorge by midmorning the next day and spirts rose noticeably as we entered Conquistador Aisle, a splendid section of canyon capped by the pine-clad peninsula of the Powell Plateau and incised in the roseate sandstones and limestones.

But we were all a little weary by now, and manifesting it in individual ways. Fred's was revealed at mealtime. Somewhere not far below Phantom Ranch he had made our hundredth meal on the river, and that routine was understandably becoming tedious. Within thirty minutes of our tying up for the day he always had soup and tea or coffee ready and the rest of a full dinner no more than ten minutes behind it. His pace never slowed, but we could tell what kind of day he'd had by the spiciness of the food. As the days went on, chili powder, garlic, oregano, onion salt, and pepper became constant constituents of salad, vegetables, and meat, until one or two passengers with nonwestern tastes asked if they could serve themselves before the "hot stuff" was added. "Oh, did I make something too hot?" Fred asked mock-incredulously, and when answered affirmatively, acknowledged with an even more incredulous "mmm!"

Little group crosscurrents were at work, too, among all of us. For the first time in several years, I had let the passengers reboard the same boat rather than having them ride with a different oarsman every day, and cliques had developed. One of the oarsmen turned out to be a Diluvianist, believing that Grand Canyon was created by the Great Flood five or six thousand years ago. The orthodox amateur geologists among us scorned that view, and so its holder. There was some kind of issue among the women about seniority on the river, and an attempt to develop a "pecking order" among them. The Campsite Game was pushed whenever the river and the daily schedule allowed leeway for it, and I had begun thinking of the *Camscott* as the *"Campscoff."*

The boats were tired, too. Except for the *Camscott*, new and all aluminum, their chines were dented and scraped, and leaking into the watertight compartments. The morning after Crystal, after we had spent the evening hours repairing one of the boats by twilight, and then flashlight, her oarsman made a careless run of the day's first rapid, over a barely exposed rock, and holed her so badly that bailing barely kept her afloat. Annoyed, I had the boats land just below, and we again emptied her of supplies, duffel, and water, and made more repairs.

But we ran "Deubie" — Deubendorff Rapid — and with seven boats making seven good runs of a difficult channel, a lot of the exhilaration of river-running came back again. Our human games were largely put aside in favor of enjoyment of the remaining miles. We showered under the pelting cold water that drops down a hundred-foot feathery cascade at Deer Creek Falls, and afterward hiked the steep trail that climbs above and behind the falls to its source in a spring-fed creek of Surprise Valley. There we sprawled for an hour in the shade on the flat limestone floor, listening to the rattle of the water on its way to the falls, and to the beguiling sound of the canyon wrens' cascading warble. That night Fred spontaneously added one of our favorite courses to the meal — Navajo fry bread. When we lined Upset Rapid the next day, and Lava Falls two days after that, we did them cheerfully and well.

Two nights before we reached Diamond Creek, a heavy rainstorm formed downriver. Malcolm, Priscilla, and I watched it from a shallow cave where we had taken refuge against the rain that soon came up the river, soaking our camp. To the south, the clouds were steel gray at first, then they became as heavy and black as the schists of the Granite Gorge, and the lightning so strong that its stabs hurt our eyes. Almost certainly, heavy rain was falling on Diamond and Peach Springs

*Below the Granite Gorge, into the roseate sand-
stones and limestones again, and nearing journey's end*

Creeks, flashflooding the little road that comes to the river there. Getting the boats out would be impossible until bulldozers had reworked the creek bed the road follows for most of its twenty miles from the Hualapai Indian settlement of Peach Springs. We would have to leave the boats and walk out, carrying as much of our gear as we could.

It seemed as though getting out of the canyon would be as hard as anything Major Powell and his men ever did. We started up the creek bed at midmorning, carrying essentials: exposed film, valuables, and some duffel, wading in and out of ankle-deep water still running down the flood-gutted canyon floor. Where the road crisscrossed the stream, the storm had cut it, leaving two- and three-foot drop-offs. Aprons of gravel and boulders had been flood-thrown everywhere, and there was no easy route.

Then, when we'd gone almost two miles, we heard the unmistakable sound of a Jeep, geared down low, engine revving high. In a few hundred yards we saw it working toward us, and behind it a Ford Bronco, picking their way between the biggest rocks, ramming their way over the smaller ones. We stopped walking and let them come to us. The Jeep belonged to our friends, the Byers, from Peach Springs, and Joan was driving the Bronco. There was enough room in and on the two vehicles to shuttle us the remaining six miles, to where the road was undamaged and cars were waiting.

At Peach Springs, after everyone else had left for Flagstaff, Fred and I made our way into the nondescript little roadside café, eager for something cold and wet. It was deserted except for a Hualapai girl and her boyfriend, who were playing pool in a side room. We sat down and waited. The thought of a milk shake was almost an obsession by now. The pool game was going slowly, and it was a half hour before the girl finished it and brought us water and a little shaved ice in scuffed plastic glasses. Making the milk shakes took another fifteen minutes; then she went back into the pool room to start another game. We emptied our glasses almost before she had gone, and wanted another round, but it seemed like too much trouble and so we went to the cash register and stood, and when she had finished the game she came and took our money.

When it was all over, I had to go away from the river, to ponder whether we'd done anything worth writing about, after all. Deciding on Mexico, I found a quiet beach on the Sea of Cortez, not far from where the Colorado finally goes home to the sea. It seemed an appropriate place to be; we had started our expedition near the point where the Green begins. In time, the fatigue and strain drained away, the good memories of the expedition began dominating, and the summer became mono-dimensional.

What had we done? Started on the same day in May and run in forty-three days the river that had taken the Powell expedition almost twice as long. Reoccupied some of the historic campsites, camera stations, and observation points. Lined only one-tenth of the rapids Powell's men had had to line. Extended the known boundary of an ancient lake fifty miles eastward. Introduced some new people to the river.

In the beginning it had been an exciting prospect: to retrace the river route of discovery at the century mark. But during the summer, the romance and drama seemed to have gotten lost. The symbolic departure ceremonies at Green River, Wyoming, had turned into almost endless speeches by the politicians, national, state, and local, and it was a great relief to get away, and onto the river. Up on the Green, the river was new, and we were fresh and eager, but by the time we had begun to feel the cumulative strain and fatigue, the new river with its aura of discovery was far behind. Then, returning to my desk late in the summer, to a massive accumulation of mail and clippings, I found that most of the centennial publicity had gone to the politicians, and to some others who had run "paper expeditions," having put out noisy press releases, but never having run the river.

In the end, it appeared that we hadn't done it for Powell, or posterity, or the politicians. Only for ourselves, which is the best way it could have been done — I can see that now. Discovery was the essence of the 1869 expedition, and discovery is a personal thing. The original voyage of discovery is reiterated every time someone new sees the river for the first time.

A newspaper editor, in alleging that Grand Canyon had more worth as a location for hydropower dams than as a setting for wilderness experience, once told his readers that river-running was for "only the reckless few." Despite the context of his remark and the thrust it was meant to have, he was right. Broken waters don't sing for everyone. But the solitary man or woman has been forgotten in the mushrooming madness for mass recreation. Running a river is one of the few remaining ways to compete against nature rather than against others, or against society. It's a wonderful change, a wonderful struggle, because the river lets you know immediately whether you've won or lost. In the battles of day-to-day life, one can't always recognize one's wins and losses. But the successful run of each rapid is a clear-cut victory all in itself, and the run of a whole river reiterates all of the victories along the way.

Yes — I'll teach my sons to love River. Man has hurt the Green, and the Colorado, and he will undoubtedly hurt them again. But now we have started to think about saving some of our rivers. And a pair of oars is still a good way to make music.

INDEX

Index

Index

Index